Green Energy and Technology

Climate change, environmental impact and the limited natural resources urge scientific research and novel technical solutions. The monograph series Green Energy and Technology serves as a publishing platform for scientific and technological approaches to "green"—i.e. environmentally friendly and sustainable—technologies. While a focus lies on energy and power supply, it also covers "green" solutions in industrial engineering and engineering design. Green Energy and Technology addresses researchers, advanced students, technical consultants as well as decision makers in industries and politics. Hence, the level of presentation spans from instructional to highly technical.

Indexed in Scopus.

Indexed in Ei Compendex.

Lin Chen
Editor

Trends in Clean Energy Research

Selected Papers from the 9th International Conference on Advances on Clean Energy Research (ICACER 2024)

Springer

Editor
Lin Chen
Institute of Engineering Thermophysics
Chinese Academy of Sciences & University
of Chinese Academy of Sciences, China
Beijing, China

ISSN 1865-3529 ISSN 1865-3537 (electronic)
Green Energy and Technology
ISBN 978-3-031-67986-5 ISBN 978-3-031-67987-2 (eBook)
https://doi.org/10.1007/978-3-031-67987-2

This Springer imprint is published by the registered company Springer Nature Switzerland AG
The registered company address is: Gewerbestrasse 11, 6330 Cham, Switzerland

Contents

Part I
Solar Photovoltaic Power Generation and Solar Energy Applications

Installations of Solar PV to Prevent CO_2 Emissions for Schools in Hampshire

Berna Yigit and Tugba Dincbas

1 Introduction

Modern civilization is built on a complicated network of social and economic activity; commerce, transportation, and entertainment are all interwoven and contribute to both sustaining and enhancing our way of life [1]. Nonetheless, there have been notable impacts on the environment and substantial strains on natural resources. A large portion of this is caused by our high-consumption lifestyles, which demand large quantities of resources and severely tax the ecosystem [2]. For example, more than 75% of the increase in atmospheric CO_2 since the pre-industrial level is attributed to CO_2 emissions from the fossil fuels that are burned and the manufacturing of cement [3].

It is known that solar energy, one of the renewable energy sources, is also used to reduce CO_2. Solar photovoltaics (PV) has emerged as a significant device for reducing CO_2 emissions globally [4]. In the previous 10 years, installed PV capacity has increased more than eight times globally, and in 2022, it will supply around 12.8% of the world's total power needs [5]. Additionally, about 20% of the emissions produced by the public sector are from school buildings. Moreover, between 75% and 80% of school in EU utilize traditional energy sources and are extremely energy inefficient; this indicates that need to decarbonize schools [6]. In this study, it was investigated how establishing a PV farm to meet energy needs would mitigate CO_2 emission. The determination of the PV farm's lifetime was disputed, and it depended on the module supplier's preference (especially on the

B. Yigit (✉)
Bursa Uludag University, Bursa, Türkiye
e-mail: berna.yigit@iklim.gov.tr

T. Dincbas
Directorate of Climate Change, Ankara, Türkiye

L. Chen (ed.), *Trends in Clean Energy Research*, Green Energy and Technology,
https://doi.org/10.1007/978-3-031-67987-2_1

country of manufacturing). The estimated CO_2 mitigation of such a system was investigated when considering the decarbonization roadmap for the UK grid.

2 Methodology

2.1 Heat Demand Calculation

Four hundred twenty-eight school buildings in Hampshire and two schools that have the lowest and highest U-values were selected, and analyses were made for one classroom.

An application for simulating transient systems with a modular framework is TRNSYS 18 [7]. In the study, the software was utilized to simulate the school building. Using the material information for the walls, floors, and roof of two chosen schools, a new example model was made. Using the tool, the U-value and details on the building materials used in two schools were used to compute the heat demand. The TRNSYS 18 software was used to simulate using weather data [8] from London Gatwick.

2.2 Life Cycle Assessment for Solar Panels

The PVs in the Hampshire region and the amount of power generated by these panels were identified in the study. Furthermore, through a literature analysis, the CO_2 emissions that arise from photovoltaic cell manufacture and the associated production costs are examined. A complete life cycle study of the PV production process has been conducted for this purpose. Thus, if a PV with a 1 kWp capacity is created, the amount of CO_2 emission avoided by installing a solar panel was calculated using the national grid's CO_2 intensity data.

3 Results and Discussion

3.1 Heat Demand Calculation of Selected Schools

Figure 1 shows the TRNSYS 18 program output and Table 1 shows the heat demand calculations for two selected schools. In addition, heating negative cooling has positive value in TRNSYS output. As it is seen in Fig. 1, summertime appears to bring with it a greater need for heat. This has to do with the program that displays summertime warmth as negative value.

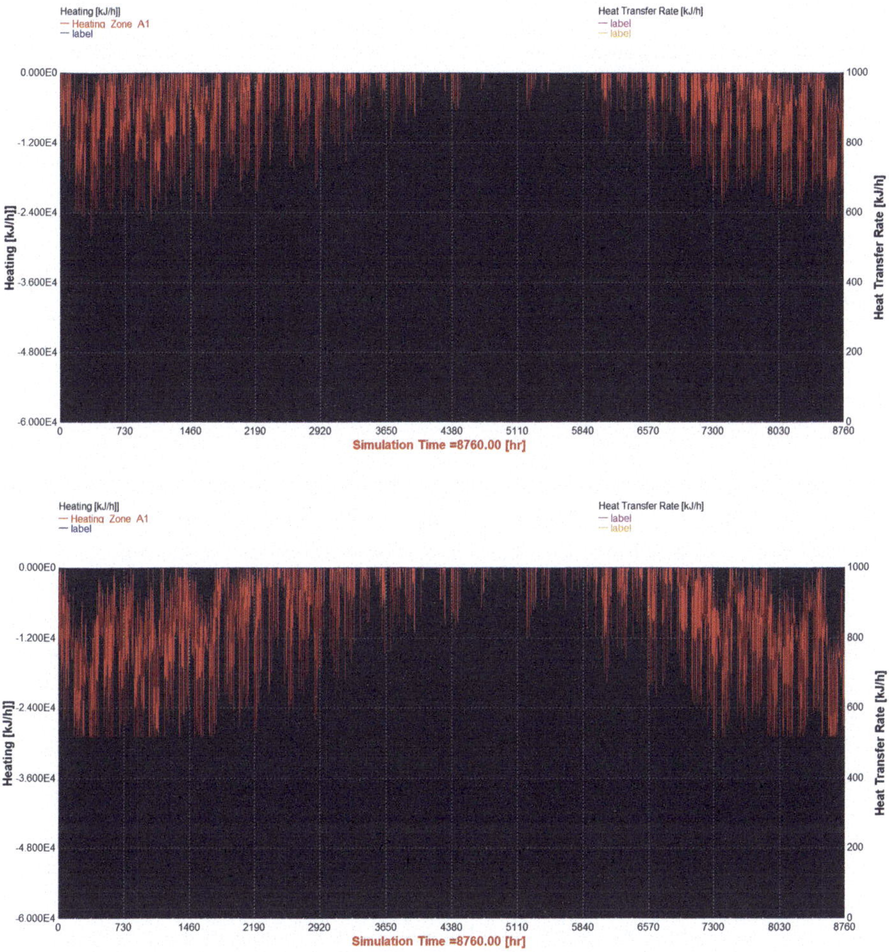

Fig. 1 Heat demand pilots of selected schools

The total useful floor area is 1703 m^2 of School 1 (SCH1), and the total energy demand was estimated to be 82 kWh/m^2 $\times$ 1073 m^2 = 87,986 kWh per year. The total useful floor area is 2160 m^2 of School 2 (SCH2), and the total energy demand was estimated to be approximately 177 kWh/m^2 $\times$ 2160 m^2 = 382,320 kWh per year.

The use of different materials in the building construction are directly related to heat demand and CO_2 emissions. Better building materials are utilized in schools and less heat is lost when the U-value is lower [10].

Regarding Table 1, it is observed that the heat demand in SCH1 is less than SCH2. It is well known that the materials chosen for buildings affect variations in heat demand [11]. The windows are single-glazed and face south for the

Table 1 Heat demand calculation [9]

SCH1 (per year/per classroom)						
	North wall	East wall	South windows	West wall	Roof	Ground floor
Glazing	None	None	Single	None	–	–
U-values (W/m^2.°C)	0.469	0.469	5.8	0.469	0.257	0.513
Area (m^2)	30	24	20	24	80	80
Heat demand = 82 kWh/m^2						
SCH2 (per year/per classroom)						
	North wall	East wall	South windows	West wall	Roof	Ground floor
Glazing	None	None	Single	None	–	–
U-values (W/m^2 °C)	1.73	1.73	5.8	1.73	1.397	0.282
Area (m^2)	30	24	20	24	80	80
Heat demand = 177 kWh/m^2						

two model schools that have been chosen. These materials' heat transmittance is directly correlated with their U-value. In other words, improved and altered building materials are needed to lower the heat demand determined by the U-value.

The best wall material (U-value = 0.469 W/m^2 °C) and the best roof material (U-value = 0.257 W/m^2 °C) was seen at SCH1. However, the ground floor material is better in SCH2 (SCH1 U-value = 0.513 W/m^2 °C and SCH2 U-value = 0.282 W/m^2 °C). It was noted that the school requiring the highest heat was SCH1, because of the better construction material (although the ground floor material of SCH1 is worse compared to SCH2). It was pointed out that there was no impact on the need for heat because of the ground floor material since there was not much differentiation between both school's U-values.

Similarly, TRNSYS was used by Rasheed et al. [11] to examine the U-value of a variety of materials, including polyvinyl chloride (PVC, 4% lower than PE), polycarbonate (PC, 9% lower than PE), and polyethylene (PE) and to ascertain the impact of indoor-outdoor temperature changes on the U-values of these materials. In the study, it was observed that U-values also changed as the materials used changed. Additionally, a 34% decrease in heat loss was seen when double glazing was used.

3.2 Life Cycle Assessment of Solar Panels in Hampshire

Life Cycle Assessment for PV Energy technology comparisons can be facilitated and "cradle-to-grave" environmental burdens can be identified with the help of LCA. The life cycle phases of photovoltaic power plants encompass the manufacture of materials and modules, in addition to the rates of greenhouse gas emissions from each stage [12].

Variations in inventory data (such as material consumption, electricity mixes, and energy requirements) in addition to differences in approach parameters (such as solar radiation, lifetime, module efficiency, and performance ratio) and system

Table 2 GHG emissions for solar panels

Location	GHG emissions (g CO_2/kWh)	Location	GHG emissions (g CO_2/kWh)
Japan [13]	53.4	Brazil [17]	30
Italy [14]	26.4	Singapore [18]	29.2
Netherlands [15]	30	Germany [19]	28.5
USA [16]	72.4	Bangladesh [20]	72.4

Table 3 Performance of grid-connected PV

Provided inputs	
PV technology	Crystalline silicon
PV installed (kWp)	1
System loss (%)	14
Simulation outputs	
Azimuth angle (°)	−2
South facing slope angle (°)	39
Yearly in-plane irradiation (kWh/m^2)	33.17
Yearly PV energy production (kWh)	1004.15

boundaries (such as including or excluding balance of system (BOS), transport, and end-of-life treatment) can all be used to explain differences in g CO_2/kWh values in solar PV production. Table 2 summarizes the greenhouse gas emissions for solar panels based on several research investigations.

Table 2 illustrates that Italy's PV module manufacturing has less of an impact on CO_2 emissions than the USA and that the EU's module output is either somewhat lower or comparable to that of Italy. Sc-Si PV panels were employed in the research of Luo et al. [18] and Stamford et al. [19]. The CO_2 emissions in the studies by Luo et al. [18] and Stamford et al. [19] were 29.2 g CO_2/kWh and 28.5 g CO_2/kWh, respectively. These findings imply that comparable emissions will occur from the use of comparable materials in PV manufacturing. Table 2 illustrates how CO2 emissions in PV manufacturing differ depending on the producing nation.

Life Cycle Assessment of PV for 1 kWp Capacity The performance of grid-connected PV systems with a 1 kWp capacity (small scale) was computed in this section. For the computation, the Photovoltaic Geographic Information System [21] was utilized. Table 3 displays the results of such computations.

The projected yearly power output from the PV farm is 1 kWp (1000 kWh). An overview of the averted CO_2 computation may be seen below. Values from the Department for Business, Energy, and Industrial Strategy [22] were used in the computations. Table 4 displays avoided CO_2 values together with their projected future values. These numbers indicate future projection. The national grid CO_2 intensity [23] was used to calculate avoided CO_2 projections, and these values are 0.2 kg CO_2/kWh in 2020 and 0.006 kg CO_2/kWh in 2050: 0.2 kg CO_2/kWh $\times$ 1000 kWh $=$ 200 kg CO_2.

Table 4 Avoided CO_2 values and future CO_2 projections

Electricity grid year	kg CO_2/kWh	kg CO_2	Electricity grid year	kg CO_2/kWh	kg CO_2
2020	0.2	200	2036	0.019	19
2021	0.138	138	2037	0.017	17
2022	0.129	129	2038	0.017	17
2023	0.124	124	2039	0.016	16
2024	0.135	135	2040	0.014	14
2025	0.114	114	2041	0.012	12
2026	0.084	84	2042	0.011	11
2027	0.070	70	2043	0.011	11
2028	0.065	65	2044	0.010	10
2029	0.060	60	2045	0.009	9
2030	0.048	48	2046	0.008	8
2031	0.038	38	2047	0.007	7
2032	0.033	33	2048	0.007	7
2033	0.029	29	2049	0.006	6
2034	0.026	26	2050	0.006	6
2035	0.023	23			
Total					**1443**

It is predicted that the installed PV would last for 30 years. Table 4 illustrates that when the PV lifespan shortens, less CO_2 is avoided. Based on the computations, it is estimated that a 1 kWp insulation will stop 1443 kg of CO_2 emissions in 30 years.

The study also contrasted solar panels made in China and Europe, discussing which one created less CO_2.

When the average grid CO_2 intensity is taken into account over a 30-year period, PV saves 1443 kg CO_2, and then the net CO_2 saving is 1443 kg CO_2–2333 kg $CO_2 = -890$ kg CO_2. China is producing 2333 kg CO_2/kWp CO_2 for mono-Si PV [24].

When taking into account the average grid CO_2 intensity over a 30-year period, PV saves 1443 kg CO_2; hence, the net CO_2 saving is 1443 kg CO_2–1467 kg $CO_2 = -24$ kg CO_2. The EU is creating 1467 kg CO_2/kWp CO_2 for mono-Si PV [24].

The same comprehensive information on the panel's lifetime and the link for the following 30 years was given by Fukurozaki et al. [17]. They highlighted the pace of CO_2 emissions, global warming potential, and cumulative energy use in their study. The LCA of the PV system has an 843 kg CO_2 potential for global warming. A maximum of 159 kg of CO_2 was released during the panel production phase, while 151 kg of CO_2 was produced during the building phase.

4 Conclusion

In the study, since reducing CO_2 emissions is critical to achieving net zero emission targets, the study investigating whether CO_2 emissions can be prevented in case of establishing a PV farm was discussed.

First of all, heat demand calculation was made, and the heat demand has been found to be 82 kWh/per year for SCH1 and 177 kWh/per year for SCH2. It is found that changing construction materials, for example, using double glaze instead of single glaze, may be a more effective method to reduce the heat demand in buildings.

The second calculation involved the installation of solar panels and the total amount of CO_2 emissions that will be avoided until 2050. The research then made future projections. The study examined two distinct PV-generating areas. For every kWp of photovoltaic power, PV farms constructed in the EU avoided −24 kg of CO_2 compared to −890 kg for those made in China. Nevertheless, the computations and investigations have shown that CO_2 cannot be conserved; on the contrary, the construction of PV farms would result in a rise in CO_2 production. However, by producing PV modules and PV components with more advanced technology and by using more clean energy sources as main energy sources, the CO_2 footprint of PV systems might be decreased.

Therefore, the results show that the PV module will not save CO_2 for both EU and China in Hampshire. On the contrary, PV modules produce additional CO_2 emission over 30 years. It has been observed that PV farms will not be effective in meeting the energy needs of the UK due to the low number of sunny days. In the study, $+CO_2$ cannot be obtained to meet the heat needs of even the school with the lowest U-value, and solar PV production is calculated as $-CO_2$. It has been understood that in future studies, the life cycle of CO_2 emissions should be studied by changing the construction materials used in buildings.

References

1. Zhu, Z., Wang, Y., Yuan, M., Zhang, R., Chen, Y., Lou, G., Sun, Y.: Energy saving and CO_2 reduction schemes for families with the household PV-BES-EV system. Energy Build. **288**, 113007 (2023)
2. Hammond, G., Jones, C., Lowrie, F., Peter, T. (eds.): Embodied CO_2, the Inventory of CO_2 and Energy (ICE). University of Bath (2011)
3. Solomon, S., Qin, D., Manning, M., Chen, Z., Marquis, M. Averyt, K.B., Tignor, M., Miller H.L. (eds.): The Physical Science Basis: Contribution of Working Group to the Fourth Assessment Report of the Intergovernmental Panel on Climate Change, IPCC Fourth Assessment Report (AR4) (2007)
4. Jiang, H., Lu, N., Wang, X.: Assessing CO_2 reduction potential of rooftop PV in China through remote sensing data-driven simulations. Sustain. For. **15**(4), 3380 (2023)
5. International Energy Agency (IEA).: https://www.iea.org/energy-system/renewables/solar-pv (2023). Accessed 19 Oct 2023
6. Dixson-Declève, S., Spence-Jackson, H.: Social innovation for Decarbonisation: the atlas school project. In: Osburg, T., Schmidpeter, R. (eds.) Social Innovation. CSR, Sustainability, Ethics & Governance. Springer, Berlin, Heidelberg (2013)

7. TRNSYS 18 (Transient Systems Simulation Program).: https://sel.me.wisc.edu/trnsys/ (2023). Accessed 8 Oct 2023
8. Energy Plus, Weather Data by Location-London Gatwick 037760 (IWEC).: https://www.google.com/search?q=Energy+Plus%2C+Weather+Data+by+Location-London+Gatwick+037760+(IWEC)&rlz=1C1GCEU_enIN1080IN1080&oq=Energy+Plus%2C+Weather+Data+by+Location-London+Gatwick+037760+(IWEC)&gs_lcrp=EgZjaHJvbWUyBggAEEUYOdIBBzMyMGowajSoAgCwAgE&sourceid=chrome&ie=UTF-8. Accessed 9 Aug 2021
9. Yigit, B.: Whole Life Carbon Decision Making Within Hampshire County Council Schools. University of Southampton (2021)
10. Rabiu, A., Na, W., Akpenpuun, T.D., Rasheed, A., Adesanya, M.A., Ogunlowo, O.Q., Kim, H.T., Lee, H.W.: Determination of overall heat transfer coefficient of greenhouse energy-saving screens using TRNSYS and hotbox methods. Biosyst. Eng. **217**, 83–10 (2022)
11. Rasheed, A., Lee, J.W., Lee, H.W.: Development of a model to calculate the overall heat transfer coefficient of greenhouse covers. Span. J. Agric. Res. **15**(4), e0208 (2018)
12. DUKES (Digest of UK Energy Statistics).: Chapter 5: Electricity, https://assets.publishing.service.gov.uk/government/uploads/system/uploads/attach-ment_data/file/1006701/DUKES_2021_Chapter_5_Electricity.pdf. Accessed 10 Aug 2023
13. Hondo, H.: Life cycle GHG emission analysis of power generation systems: Japanese case. Energy. **30**(11–12), 2042–2056 (2005)
14. Battisti, R., Corrado, A.: Evaluation of technical improvements of photovoltaic systems through life cycle assessment methodology. Energy. **30**(7), 952–967 (2005)
15. Alsema, E.A., Wild-Scholten, M.J.: Environmental impacts of crystalline silicon photovoltaic module production. In: 13th CIRP International Conference on Life Cycle Engineering, pp. 103–108, Leuven (2006)
16. Pacca, S., Sivaraman, D., Keoleian, G.A.: Parameters affecting the life cycle performance of PV technologies and systems. Energy Policy. **35**(6), 3316–3326 (2007)
17. Fukurozaki, S., Zilles, R., Sauer, I.: Energy payback time and CO_2 emissions of 1,2 kWp photovoltaic roof-top system in Brazil. Int. J. Smart Grid Clean Energy. **2**, 164–169 (2013)
18. Luo, W., Khoo, Y.S., Kumar, A., Low, J.S.C., Li, Y., Tan, Y.S., Wang, Y., Aberle, A.G., Ramakrishna, S.: A comparative life-cycle assessment of photovoltaic electricity generation in Singapore by multicrystalline silicon technologies. Sol. Energy Mater. Sol. Cells. **174**, 157–162 (2018)
19. Stamford, L., Azapagic, A.: Environmental impacts of photovoltaics: the effects of technological improvements and transfer of manufacturing from Europe to China. Energ. Technol. **6**, 1148–1160 (2018)
20. Rahman, M.M., Alam, C.S., Ahsan, T.M.: A life cycle assessment model for quantification of environmental footprints of a 3,6 kWp photovoltaic system in Bangladesh. Int J Renew Energy Dev. **8**(2), 113–118 (2018)
21. Wang, Z., Fan, W.: Economic and environmental impacts of photovoltaic power with the declining subsidy rate in China. Environ. Impact Assess. Rev. **87**, 106535 (2021)
22. Photovoltaic Geographical Information System, Performance of Grid Connected PV.: https://re.jrc.ec.europa.eu/pvg_tools/en/#PVP (2023). Accessed 9 Oct 2023
23. Department for Business, Energy, and Industrial Strategy.: https://www.gov.uk/govern-ment/organisations/department-for-business-energy-and-industrial-strategy (2023). Accessed 11 Oct 2023
24. National Grid, CO_2 Intensity.: https://CO2intensity.org.uk/ (2023). Accessed 10 Aug 2023

Energy Evaluation of Photovoltaic Integration in Student Building with Different PV Technologies

Juan Lata-García, Pablo Parra, Gary Ampuño, and Wellington Maliza

1 Introduction

In recent decades, with the change in society's consumption habits and new technologies, energy consumption in buildings is around 35% of the total final energy consumption, 51% of electricity consumption worldwide, and 37% of the polluting CO_2 emissions that cause the greenhouse effect. With this alarming data, governments and research entities are looking for ways to save energy and look for integrated generation sources in buildings [1, 2].

Photovoltaic energy systems integrated in buildings include many solutions depending on their application. In recent decades, transparent BIPV (building-integrated photovoltaics) has been developed and implemented [3]. One of the characteristics of this technology is that it uses natural light and converts it into lighting. The manufacturer Onyx announces the development of the first transparent architectural glass to be integrated into the S building that generates photovoltaic energy, and it filters the entry of heat into the building [4]. Photovoltaic systems are currently one of the most dynamically developed technological careers [5]. Looking at the annual data on new emerging installations in Poland, you can see that the trend is doubling. It means the fact that the number of new installations almost doubles with each passing year. In buildings, the available rooftop space for solar photovoltaics (PV) may not be sufficient to make a major contribution to the building's renewable energy supply.

J. Lata-García (✉) · P. Parra · G. Ampuño
Universidad Politécnica Salesiana- GIPI, Guayaquil, Ecuador
e-mail: jlatag@ups.edu.ec

W. Maliza
Universidad Politécnica Salesiana - GIPI, Guayaquil, Ecuador

L. Chen (ed.), *Trends in Clean Energy Research*, Green Energy and Technology,
https://doi.org/10.1007/978-3-031-67987-2_2

Integration into buildings beyond demonstration or educational projects should be encouraged. More than providing electrical power, BIPV can enhance and satisfy a building's image. BIPV makes a statement about innovative architecture as well as engineering design [6].

Research shows that mid- and high-rise buildings are ideal for incorporating photovoltaics (PV) into the façade and roof due to new building photovoltaic integration (BIPV) applications. Photovoltaic integration generates electricity to the building and improves the thermal performance of the envelope; however, there is very little research on BIPV-ventilated façades in the literature [3], especially in the Persian Gulf regions. Due to the harshness of Dubai's hot climate, the authors conducted comparative experimental research on the energy performance of BIPV-ventilated façades, where they monitor the energy performance and temperatures of copper, indium gallium (CIGS), and silicon selenide modules on the south, east, and west façades for an entire year considering the orientation, temperature, and tilt angle of the PV modules. The results show that silicon-based modules have a higher performance in all tests with an average difference of 13.6%, and west-facing modules had higher temperature values than east-facing ones [7].

Entering the room in winter significantly reduces the production of electricity.

The use of solar panels based on crystalline silicon for electricity generation has been widely studied and used around the world. Being one of the leading nonconventional renewable energy technologies within the field, the integration of this type of panels in buildings is complex due to the type of angle of inclination and orientation it has to generate at maximum power; therefore, they cannot replace the building envelope or be under total or partial shadows [8, 9].

2 Types of Façades in Buildings

Photovoltaic façades are an innovative solution that generates especially electrical energy in a sustainable way. They can be classified into three types: (A) opaque dynamic façade (walls, wall panels, etc.), (B) transparent dynamic façade (smart windows, double-skin surfaces, air chambers, etc.), and (C) semitransparent dynamic façade (shutters). Each type has different behaviors and mechanisms integrated into buildings [1].

2.1 Dynamic Opaque Façade

The main function of this type of façade is to control the transport of heat, which is embedded in its natural static behavior. Opaque surfaces can adversely affect energy consumption due to opaque behavior. In reference [2] the authors confirm that a wall with a lower U-value has a lower annual energy consumption in the building; however, this may not always be correct. The usual approach to adaptive façade design relies on detailed simulations of specific façade components. Two technology-oriented approaches tend to be incompatible with the early-stage design process and rarely make a conscious analysis of potential climate resources, which could result in suboptimal façade adaptation strategies. Users only define the location, geometry, and location of the façade [10].

2.2 *Transparent Dynamic Façade*

This type of technology is used to control the energy needed for cooling, heating, or electricity consumption. The characterization and analysis of transparent dynamic façades is more complex due to the type of visual transmission they possess. The façades use different glazing materials and technologies in order to minimize heat loss, radiation reflectance, natural light levels, and, above all, visibility. Smart glasses use smart materials such as phase change material (PCM), double-skin façades, integrated façades, and multifunctional façades which are the most promising clear glasses for marketing.

There are two types of technology in glazing, passive and active. The former do not have the supply of any source of electrical energy. The active control is decoupled, the installation is simple, and the maintenance is low as there are no moving parts. Environmental conditions must be taken into account to avoid reducing transparency and losing natural light.

Manufacturers have developed different smart window technologies. Thermochromic adapts temperature changes to visual and optical properties, electrochromic adapts the visual and optical properties to the applied voltage, and photovoltaic glazing converts and collects photovoltaic energy and distributes it among the building's charging needs. Currently, several smart windows have been introduced in the market, where manufacturers have improved electrochromic (EC) glazing with innovative advanced technologies.

2.3 *Semitransparent Dynamic Façade*

This semitransparent technology has shading devices that are primarily controlled by users in patterns. Shades are semitransparent surfaces that help reduce solar gain through walls or windows that diffuse solar radiation. The thermal mass of shadows is usually small and is not normally involved in overall heat transfer or considered in thermal calculations [5].

2.4 Efficiency Modeling

The performance of photovoltaic systems is mainly based on the amount of solar irradiance measured in (W/m^2) by means of other components. It is transformed into electricity. By means of Eq. 1, the efficiency of the photovoltaic system is calculated; in the equation, the power of the panel (DC), the global solar irradiance, and the air mass intervene:

$$\rho PV = \frac{P_{DC}}{G_{POA} \times A_m} \tag{1}$$

On the other hand, to calculate the efficiency of the entire system, all the loads in the alternating current divided by the product of the global solar radiation by the air mass are considered:

$$\rho all = \frac{P_{AC}}{G_{POA} \times A_m} \tag{2}$$

3 BIPV Photovoltaic Integration

Building-integrated photovoltaics (BIPV) is gaining ground in the global market every day. Governments adopt demanding regulations and legislations related to energy efficiency to stop climate change and become energy independent. Buildings are expected to generate the energy they consume when reaching the zero-energy construction scenario. Integrating solar panels into buildings is a complex task due to the particularity of angles that the structure has, the aesthetics of the installation, the performance of the system, the long-term reliability of the technology, etc. Few software on the market specialize in BIPV technology. The PVsite program [11] solves some of the problems by seeking to demonstrate an ambitious portfolio of solar technologies and systems integrated into construction, giving a forceful and reliable response to the market requirements identified by the industrial members of the consortium in their daily activity.

3.1 Building Under Study

The Salesian Polytechnic University is a center of higher education in Ecuador. One of its three campuses is in Guayaquil. Block F of the centennial campus is used for the student library which has the following coordinates, −2.2201494 and −79,886, the building has three floors, and the front is covered by glass to allow natural light to pass through. The dimensions are 40 m long by 7.6 m wide as shown in Fig. 1 [12]. The building is used daily by up to 2000 students. At certain times, the influx of people is low; however, the energy consumption is similar.

Fig. 1 Building under study of the Salesian Polytechnic University

3.2 *Meteorological Data*

The solar radiation that reaches the building on the façade has the highest values during the months of March to August, reaching direct irradiance up to 14.58 kWh/m2 per month. In August, the global irradiance is 56.4 kwh/m2. These data are important for the simulation of the operation of the photovoltaic integration.

3.3 *Photovoltaic Modules Used in Simulation*

This work aims to energetically evaluate two types of photovoltaic technologies that are used in a building. The first type of solar panel is a transparent photovoltaic glass where it provides energy to the building and filters the entry of heat inside. The generation depends on the type of technology used in the glass along with the irradiation emitted by the sun. The BIPV module Onyx Solar X6 glass semitransparent type with hidden bars is applied as a second skin which has 5″ c-Si solar cells with 6 mm clear tempered glass. The rectangular semitransparent glazing has the following dimensions: 2250 × 760 × 13.8 mm. The manufacturer indicates that the glazing for thermal and acoustic insulation is double/triple. The second type of solar panel used in the simulation is crystalline silicon. The technology has been widely used to generate energy; however, the integration into buildings is complex due to the needs of location, tilt, orientation, and shading. In addition to these characteristics, it does not filter the entry of heat into the building or prevent the use of conventional materials in the construction of the building.

3.4 Simulation Procedure

To find out the power output of the grid-connected inverter of the photovoltaic glass, follow the steps below in the PVsite program. Firstly, data from modules, inverters, and weather variables of the site are entered as input data. Secondly, the photovoltaic system is calculated according to the installation conditions proposed by the designer. Loss factors were then used as a variable dependent on the system configuration. In addition, to consider the influence of shading effects on the module, the PV system and obstacles were constructed using a 3D modeling function. Finally, the data is obtained in annual, monthly, and hourly values. In the case of simulation with silicon solar panels, the PVsyst program is used. It has been designed as a powerful photovoltaic sizing tool being used by researchers, architects, and engineers. It is easy to use with quite simple procedures using commercial components. The meteorological base can be imported from different sources. The procedure is quick and simple where you can specify the desired power of the photovoltaic generator or the available area and select the photovoltaic module from the available library and the inverter. The program emerges and configures system components through a preliminary simulation. The advantage of the software is the color-coded warning messages.

4 Simulation Results

As indicated in the previous section, the first step is the import of the 3D building model or the design of the library building in various formats to be imported. The façade of the building is drawn in 3D using the Sketchup 8 program. With the selection of the project location and the respective download of the meteorological data, the flexible design of BIPV modules to be located on the façade begins. The selection of inverters and connection of the solar glasses is done manually or automatically.

4.1 Photovoltaic Glass Generation

The solar panels used are from the manufacturer Onyx X6 of 299 Wp, the maximum output voltage is 218.0 V, and the maximum current is 1.4A. The inverter that has grid connection is of the Centrosolar AG model of 5.2 kWp, the maximum input

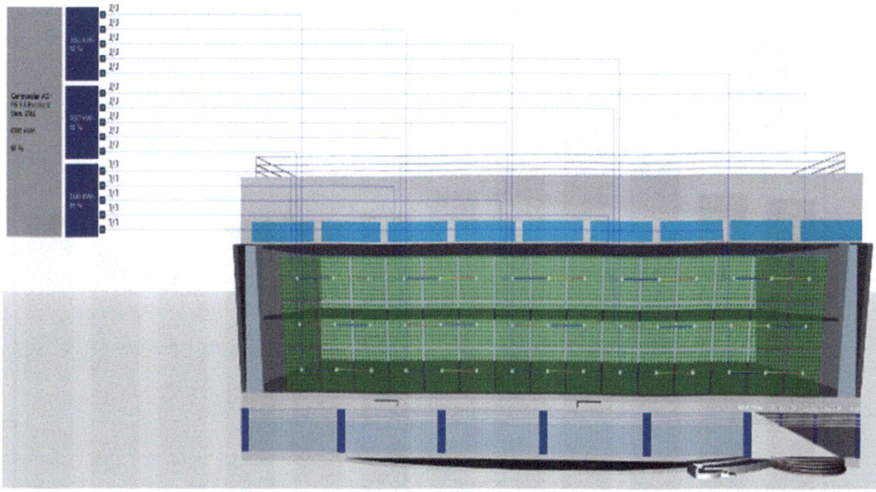

Fig. 2 Patch panels and inverter with grid connection

voltage in DC supported by the integrated controller is 950 V, and the maximum current in DC is 9A with an efficiency of 90%. With these elements, the design is composed of 28 semitransparent panels, one connecting five branches of three panels in series and a 5.25 kW inverter. The connection is shown in Fig. 2 [13].

The total power of the system is 13.5 kWp, the total efficiency is 466.8 kWh/kWp, and the output is 6281 kWh. From October to February, the electricity generation is reduced, despite being on the equator and having 12 h of sunshine per day. This occurs due to integrating solar glass into existing buildings without planning the construction of the building with respect to the position of the sun.

4.2 Generation Using Silicon Crystalline Panels

The building under study has a terrace where crystalline photovoltaic panels can be placed. The measurements to be used are 36 m × 6 m. These measurements are similar to the façade of block F, but on the other hand, the use of this technology does not prevent the entry of heat into the building; therefore, the refrigeration equipment demands more energy to achieve the comfort of the users.

For the design of the system with grid connection the PVsyst software is used. The 425 W N-type monofacial dream panel models JKM425N-54HL4-V are used, the maximum voltage is 30.7 V, the maximum current is 13.41 A, the efficiency is 21.76%, and the values are taken in STC, while the inverter to be used is from the manufacturer Sunny model Tripower STP 50, the rated power is 50 kW, and the input voltage has a range from 188 to 800 Vdc with five MPPT type inputs. The simulation results for the available area indicate that the photovoltaic system

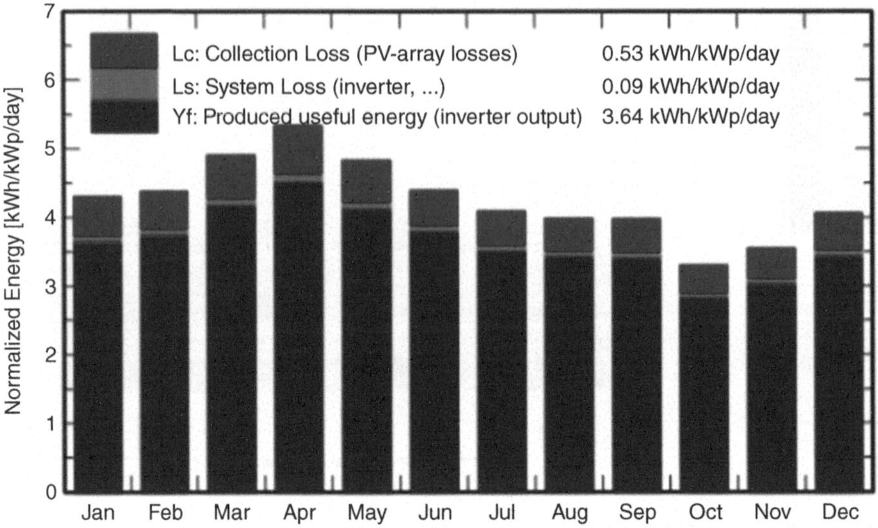

Fig. 3 Generation, unused energy and losses of the FS

is composed of 110 solar modules with a nominal power of 47 kW that can be connected in five branches in parallel. Each branch has 22 panels in series. To complete the system, a 50 kW inverter is used, the production of the system is 62.2 MWh/year as shown in Fig. 3, the specific production is 1330 kWh/kWp/year, the energy injected into the grid is 133.9 kWh/day, and the generation peak is at 12:30 with a power of 2000 W, from 18:00.

Having an energy management system is essential to take advantage of all the energy generated in the same building, the generation of the system. The energy production of the system is 3.64 kWh/kWp/day with a total generation of 62,186 kWh. The loss of the system is 0.09 kWh/kWp/day in the inverter. The highest percentage of losses is due to the temperature in which the photovoltaic panels are located. Currently, buildings seek to generate their own energy that they consume, known as zero energy buildings (CEE). The correct use of modern technologies achieves the energy and sustainable development objectives promoted by the United Nations. The production of two types of technologies to be implemented has similar surfaces, but in different locations. The first case study has an annual generation of 7090 kWh. The month with the highest production is March and with the lowest production is December with 675.7 and 469.12 kWh, respectively. In the second case study, the use of crystalline solar panels is analyzed. The annual generation of the system is 63,694 kWh. The month with the highest production is April and with the lowest production is October with 6513 kWh and 4201 kWh, respectively. Silicon photovoltaic panels have been widely used around the world and in a number of applications. Comparatively, this type of panel has a greater generation than photovoltaic glass, generating 8.9 times; however, glass technology is promising for other advantages that make it attractive when constructing a new building, avoiding the use of conventional materials and improving aesthetics and efficiency.

5 Conclusions

The integration of photovoltaics in buildings with modern technologies means energy savings by taking advantage of natural light during the day, improving insulation, and saving costs in the envelope. On the other hand, silicon-based solar panels are being used, but they need to meet some requirements for their correct operation such as location and orientation. The document analyzed the performance of two types of photovoltaic technologies that can be integrated into the building. The BIPV system is simulated in a functional building used for the library of the Salesian Polytechnic University. The first crystalline silicon technology is widely used, while the second technology is modern and very little used in the country, so the cost of the components is not easily accessible. In the near future, photovoltaic glass panels will avoid the use of conventional materials in order to be competitive in terms of generation with crystalline silicon. For a similar area of 216 m^2, the silicon panels placed on the terrace can generate 62,186 kWh per year, while the photovoltaic glass placed in the front of the building can generate 6281 kWh per year. These new systems tend to look for energy-saving solutions, which are viable when looking for zero-consumption buildings.

References

1. Shafaghat, A., Keyvanfar, A.: Dynamic façades design typologies, technologies, measurement techniques, and physical performances across thermal, optical, ventilation, and electricity generation outlooks. Renew. Sust. Energ. Rev. **167**, 112647 (2022). https://doi.org/10.1016/J.RSER.2022.112647
2. Masoso, O.T., Grobler, L.J.: A new and innovative look at anti-insulation behaviour in building energy consumption. Energ. Buildings. **40**(10), 1889–1894 (2008). https://doi.org/10.1016/J.ENBUILD.2008.04.013
3. Elhabodi, T.S., Yang, S., Parker, J., Khattak, S., He, B.J., Attia, S.: A review on BIPV-induced temperature effects on urban heat islands. Urban Clim. **50**, 101592 (2023). https://doi.org/10.1016/J.UCLIM.2023.101592
4. Xie, S., Yang, D., Zhang, J., Wu, L.: Development of semitransparent CdTe polycrystalline thin-film solar cells modified with a CuCl layer for BIPV. Sol. Energy. **267**, 112212 (2024). https://doi.org/10.1016/J.SOLENER.2023.112212
5. Runnerstrom, E.L., Llordés, A., Lounis, S.D., Milliron, D.J.: Nanostructured electrochromic smart windows: traditional materials and NIR-selective plasmonic nanocrystals. Chem. Commun. **50**(73), 10555–10572 (2014). https://doi.org/10.1039/C4CC03109A
6. Alhammadi, N., Rodriguez-Ubinas, E., Alzarouni, S., Alantali, M.: Building-integrated photovoltaics in hot climates: Experimental study of CIGS and c-Si modules in BIPV ventilated facades. Energy Convers. Manag. **274**, 116408 (2022). https://doi.org/10.1016/J.ENCONMAN.2022.116408

7. Jhumka, H., et al.: Assessing heat transfer characteristics of building envelope deployed BIPV and resultant building energy consumption in a tropical climate. Energ. Buildings. **298**, 113540 (2023). https://doi.org/10.1016/J.ENBUILD.2023.113540
8. Saretta, E., Caputo, P., Frontini, F.: A review study about energy renovation of building facades with BIPV in urban environment. Sustain. Cities Soc. **44**, 343–355 (2019). https://doi.org/10.1016/J.SCS.2018.10.002
9. Assoa, Y.B., Valencia-Caballero, D., Rico, E., Del Caño, T., Furtado, J.V.: Performance of a large size photovoltaic module for façade integration. Renew. Energy. **211**, 903–917 (2023). https://doi.org/10.1016/J.RENENE.2023.04.087
10. Habib, S., Tamoor, M., Zaka, M.A., Jia, Y.: Assessment and optimization of carport structures for photovoltaic systems: A path to sustainable energy development. Energy Convers. Manag. **295**, 117617 (2023). https://doi.org/10.1016/J.ENCONMAN.2023.117617
11. Espeche, J.M., Noris, F., Lennard, Z., Challet, S., Machado, M.: PVSITES: Building-integrated photovoltaic technologies and systems for large-scale market deployment. In: Proceedings, p. 690. MDPI (2017)
12. Juan, L.-G., Jurado, F., Pablo, P.: Optimization of energy integration renewable in a functional building. In: Industry, Innovation, and Infrastructure for Sustainable Cities and Communities, pp. 24–26 (2019). https://doi.org/10.18687/LACCEI2019.1.1.104
13. Medina, J.P., Lata-García, J.: Optimal model of a hybrid electrical system photovoltaic panel/wind turbine/battery bank, considering the feasibility of implementation in isolated areas. J Energ Storage. **36**, 102368 (2021). https://doi.org/10.1016/j.est.2021.102368

Technical-Economic Modeling of a Microgrid Incorporating Renewable Photovoltaic Energy

Gary Ampuño, Erick Guerron, and Alexander Mendez

1 Introduction

Technological advancements have enabled more people to enjoy the convenience of electricity. In 2010, 84% of the global population had access to electricity, a figure that increased to 91% by 2021. However, 15.5% of residents, particularly those in rural communities, still lack access to electricity, posing a challenge in many countries and regions in terms of electrical supply [1]. To address this issue, various organizations have implemented specific projects, such as the IEEE Smart Village, to benefit these underserved populations and empower communities in remote areas [2–3]. As of 2021, Ecuador has achieved significant coverage of electricity supply in rural and urban areas. This electricity is sourced from various generation methods, and renewable electricity generation projects, including photovoltaic energy, are being developed until 2023 to contribute to reducing greenhouse gas emissions [4, 5]. The paper is organized as follows: first, the study of loads and incident radiation is presented, followed by the economic analysis in Sect. 3. Section 4 contains an analysis of the simulation results and, finally, some conclusions are drawn.

G. Ampuño (✉) · E. Guerron · A. Mendez
Universidad Politécnica Salesiana, Cuenca, Ecuador
e-mail: gampuno@ups.edu.ec

L. Chen (ed.), *Trends in Clean Energy Research*, Green Energy and Technology,
https://doi.org/10.1007/978-3-031-67987-2_3

2 System Model

2.1 *Microgrid Design/Proposal for Building*

The electrical supply that supplies the entire load existing in the building is provided by the public electrical network, which is why, through data analysis, the design of a renewable system that serves support for possible interruptions in the operation of the building is proposed important loads in the event of a power outage, in addition to percentage reductions in the use of energy supplied by the conventional grid [6].

2.2 *Load Installed in the Building*

The load analysis is one of the main references to determine the capacity of the photovoltaic equipment and batteries. The information on the total power of the loads of each of the floors of the building is presented in Table 1, which describes the location, the power, and the percentage that the latter represents in the system.

The value of the loads presented in Table 1 includes the power value of all the existing equipment; the information was collected on the primary educational loads such as lighting and computer equipment. This is because the loads are used for teaching. The total student load power is approximately 141.55 kW, which is broken down by each of the plants as presented in Table 1. Therefore, a 150 kW solar microgrid can supply power to the above loads.

Depending on the time of use of the lighting and the PCs, it could be estimated that the total daily electrical demand is 556.2 kW/h and 1698.6 kW/h depending on the hours of operation. It should be considered that each day has a duration of 4 h, thus having a total of 12 h. Table 2 shows the demand per day and for each floor of the building.

Table 1 The value of the loads

	Installed load according to each floor of the building		Loading of educational equipment	
Description	Load [kW]	%Load	Load [kW]	%Load
Basement	95.706	14.98	4.66	4.85
Low level	30.387	4.75	1.29	4.24
Mezzanine	30.152	4.72	5.29	17.54
First floor	40.803	6.38	31.84	78.03
Second floor	188.464	29.51	13.25	7.03
Third floor	124.378	19.47	36.42	29.28
Fourth floor	128.676	20.15	48.8	37.92

Table 2 Electrical demand based on the time of use of the equipment per day

Description	Load [kW]	Load [kW/h]—12 h	Load [kW/h]—8 h	Load [kW/h]—4 h
Basement	4.66	55.92	37.28	18.64
Low level	1.29	15.48	10.32	5.16
Mezzanine	5.29	63.48	42.2	21.16
First floor	31.84	382.08	254.72	127.36
Second floor	13.25	159	106	53
Third floor	36.42	437.04	291.36	145.68
Fourth floor	48.8	585.6	390.4	195.2

2.3 Radiation Analysis

Solar radiation data were obtained from NASA's Data Access Viewer and the World Bank Group. Solar radiation has been increasing over the years, especially in cities on the Pacific coast, and due to its location in the equatorial zone, there is a warm climate all year round. Solar energy in the selected location can reach up to 430 W/m^2 of solar energy as presented in Fig. 1. The incident shortwave solar energy varies seasonally, with September being the brightest month, averaging 6.7 kWh per day, and March being the least bright month, with an average of 5.4 kWh [7].

2.4 Battery Size

Due to the magnitude of the load on the system, 48 V lithium batteries intended for solar systems were chosen, guaranteeing optimal operation.

When conducting a study of the demand for each of the floors of the building, there will be different results regarding the minimum capacity of each of the batteries. It must be considered that the selected batteries have a charge and discharge configuration of 100% and a discharge depth of 80% [8].

2.5 System Structure

The hybrid microgrid consists of the public electricity grid, the photovoltaic system, the energy storage system, and the control system. The structure of the system is presented in Fig. 2. Based on the analysis conducted in the previous points, the system would have solar panels that would provide a power of 372.8 kW and battery packs whose sum gives a total of 600 kW/h [9]. The distribution of the microgrid is shown in Table 3.

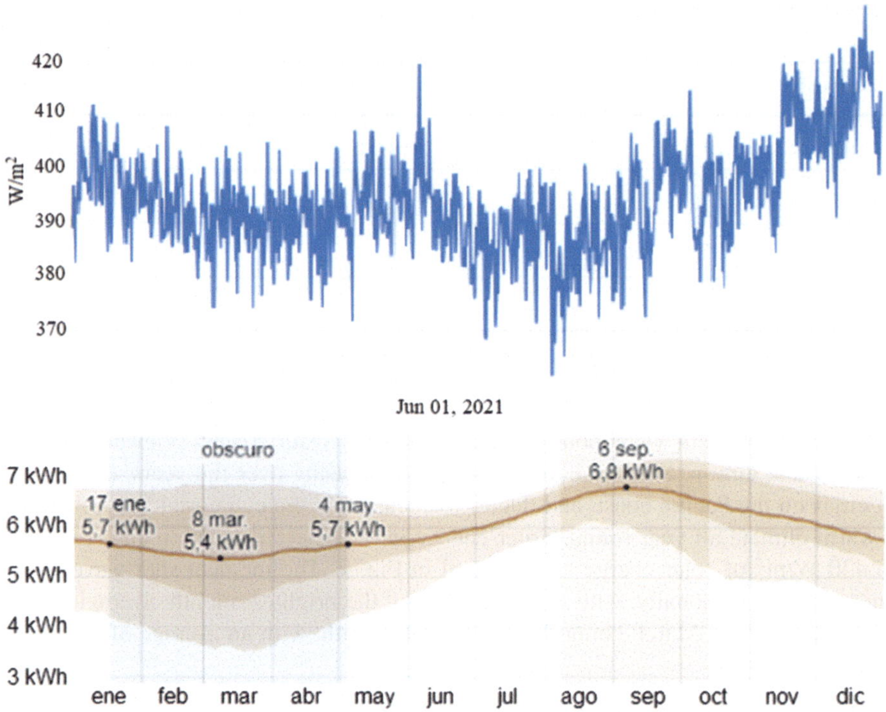

Fig. 1 (**a**) Behavior of solar radiation in the study area. (**b**) Average daily incident short ware solar energy in Guayaquil

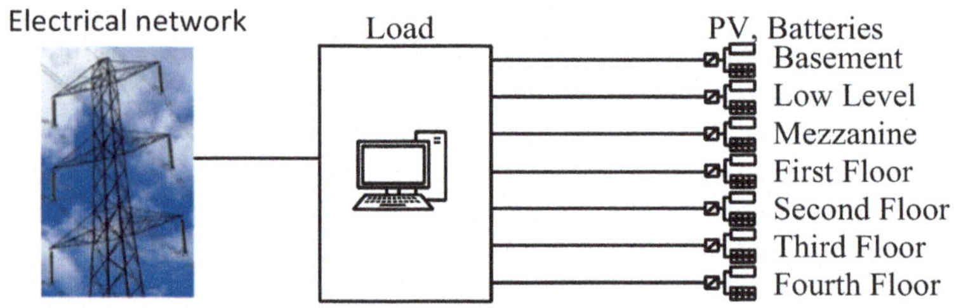

Fig. 2 Structure of the solar public grid hybrid microgrid

3 Economic Analysis

The cost of a photovoltaic system is reflected in the investment of main equipment (panels, batteries, inverters, controllers) and construction costs (tools, cabinets, cables). The costs of each component are listed in Table 4.

Table 4 shows the cost relationship of each of the elements that make up each of the systems that were designed to supply energy to each of the floors of the building.

Table 3 Elements belonging to the solar network

Description	PF	QYT	INV	QYT	BAT	QYT
Basement	430Wp	13	MIN4200TL-X	1	PT-48V100Ah	4
Low level	430Wp	5	MIC2000TL-X	1	PT-48V50Ah	3
Mezzanine	430Wp	15	MIN5000TL-X	1	PT-48V100Ah	5
First floor	430Wp	77	MAC22KTL3-XL	1	PT-48V300Ah	9
Second floor	430Wp	37	MAC10KTL3-XL	1	PT-48V250Ah	5
Third floor	430Wp	87	MAC25KTL3-XL	1	PT-48V300Ah	11
Fourth floor	430Wp	116	MAC36KTL3-XL	1	PT-48V00Ah	14

Table 4 Cost of the photovoltaic system belonging to the microgrid

Description	PF	Cost (USD)	Inv	Cost (USD)	Bat	Cost (USD)	Total
Basement	13	$190.4	1	$699	4	$3450	$16,974.85
Low level	5	$190.4	1	$419	3	$1670	$6381.25
Mezzanine	15	$190.4	1	$819	5	$3450	$20,925.75
First floor	77	$190.4	1	$2498	9	$10,200	$108,962.6
Second floor	37	$190.4	1	$2316	5	$8090	$49,812.65
Third floor	87	$190.4	1	$2498	11	$10,200	$131,267.1
Fourth floor	116	$190.4	1	$3999	14	$10,200	$168,891.2
Total	–	–	–	–	–	–	$503,215.5

As shown in the table, the costs of the panels, converters, and batteries (including connectors and cable) would represent 78–82% of the overall investment expenses.

To calculate the cost of energy generation of the photovoltaic system, the existence of investment credits, average annual investment cost, and the cost per kilowatt hour must be considered since based on these variables, the generation cost can be calculated by the following formula [10].

$$\mathrm{CPU} = \frac{\mathrm{CA}}{\mathrm{GEA}} = \frac{\mathrm{CRF}\,(i, n) * \mathrm{TC} + \mathrm{CMA}}{\mathrm{HEA} * P} \tag{1}$$

where AC denotes the annual system cost, AMC represents the annual maintenance cost, TC represents the total cost of the system investment, CRF is the capital recovery factor (calculated based on the interest rate and its period), CPU indicates the electricity cost per unit, and HEA represents the annual equivalent hours of solar generation. GEA represents the annual solar energy generation and P represents the installed photovoltaic capacity.

In Ecuador, there are banking entities that offer loans with interest ranging from 5.5% to 9.5%. Therefore, with a 30-year loan with an interest rate of 5.5%, an average of 1480 h of solar generation, and a maintenance estimate of $1200/year, the result is an average cost for each kilowatt/hour generated by the entire system would be approximately $0.077/kWh. It should be noted that the cost of power generation using gasoline generators or the construction, operation, and maintenance of a substation will be much more expensive than the use of solar energy [11].

3.1 Microgrid Optimization

During the design of the microgrid, with the help of the PVsyst program, the lowest cost was found among the feasible operations. This process, which consisted of several tests, was conducted with the purpose of obtaining the optimal quantities of equipment that would provide us with adequate energy generation to cover the largest possible electrical load by implementing a minimum amount of equipment for each system. This is because renewable energy resources have high investment costs, also considering that this type of generation has a low operating cost. The economic minimization model is defined as follows [12]:

$$\min\left(\mathrm{NPC}_t\right) = \min \sum\nolimits_{i=1}^{n} \frac{(K_{t+i} + RP_{t+i} + OM_{t+i})}{(1+r)^i} \tag{2}$$

where NPC represents the net present cost, r is the interest rate, K_{t+i} represents the initial capital cost in $t + i$ years (if $i = 0$, it means base year), and RP_{t+i} represents the replacement rate in $t + i$ years (if $i = 0$, it means base year).

4 Microgrid Operation

DigSilent Power Factory was used to analyze through simulations the effect that the implementation of photovoltaic energy would have on the electrical system belonging to the building.

4.1 Original Microgrid Operation Scheme

The system operates based on the energy supplied by the conventional public grid, connected to a three-phase transformer that reduces a voltage from 69 to 13.8 kV to travel about 2.5 km to the building where it will be connected to a new transformer to reduce from medium voltage to low voltage, i.e., it is transformed from 13.8 to 240/120 V, respectively. For this analysis, the power of all existing loads on each of the floors of the building was entered, the same as detailed in Table 1, to observe the behavior of the network to have 100% of the load in operation. This is shown in Fig. 3.

4.2 Hybrid Microgrid Operation Scheme

Once the implementation and simulation of the design of the photovoltaic systems proposed in Table 4 has been conducted, a new analysis of the operating values of the equipment is obtained as a result. In this way, the behavior of the hybrid network

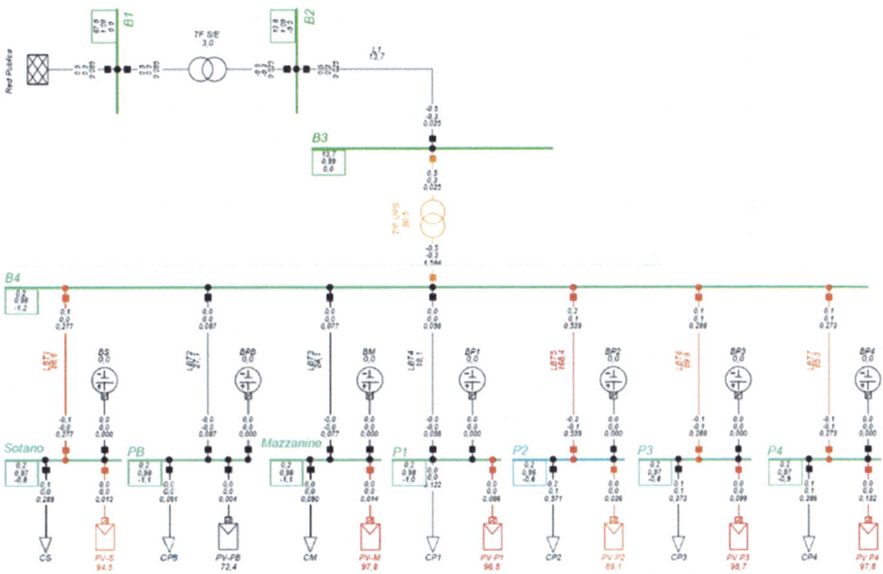

Fig. 3 Conventional electrical network with the implementation of photovoltaic energy

can be known. For the benefits of implementing solar energy to the system since, as can be seen in Fig. 4, while the solar system is operating, there will be a decrease in the use of the conventional network, which would represent a reduction in the payment for said energy when consuming a lower amount of kW/h.

5 Conclusions

The project proposes the implementation of a photovoltaic microgrid to supply a building belonging to a third-level educational institution in Ecuador. The primary objective is to cover the fundamental demands of lighting and educational loads while seeking to reduce energy consumption from the conventional power grid and avoid the need to build a substation. The building in question houses a variety of laboratories specialized in professional practices related to different engineering fields.

The system configuration comprises solar panels with a total generation capacity of 372.8 kW, a 600kWh battery bank, and hybrid converters of varying capacities: 2.1, 5.8, 6.4, 15, 33, 37.5, and 54 kW. In situations of low PV generation or when the storage system does not meet the designated loads, the conventional grid steps in to supply power. The capacity of the solar panels and batteries is determined by detailed load and solar radiation analyses, ensuring the ability to meet basic lighting needs and teaching loads. In addition, the system was designed to maintain power

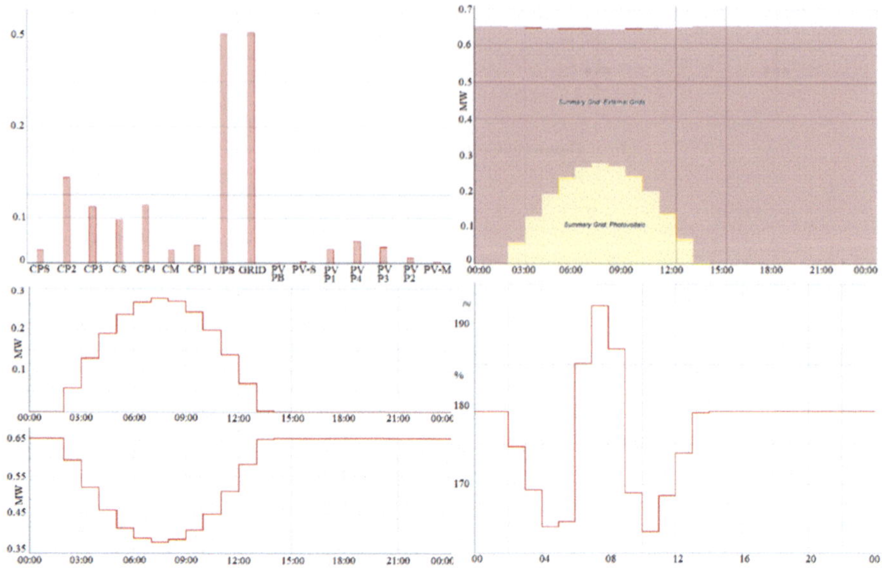

Fig. 4 (**a**) Power behavior of each of the system elements, (**b**) generation of the conventional network versus generation of the photovoltaic system, (**c**) behavior of the electrical power of the system, and (**d**) behavior of the installed load at the time of hybrid operation

supply to these loads for approximately one and a half days of university classes (equivalent to 4–6 h) in the absence of sunlight.

References

1. Noumba, I., Nguea, S.M.: Assessing the role of globalization for universal electricity access. Int. Econ. **174**(April), 180–195 (2023). https://doi.org/10.1016/j.inteco.2023.03.008
2. Prinsloo, G., Mammoli, A., Dobson, R.: Customer domain supply and load coordination: a case for smart villages and transactive control in rural off-grid microgrids. Energy. **135**, 430–441 (2017). https://doi.org/10.1016/j.energy.2017.06.106
3. World Bank: Global development indicators, electricity access, rural (% of population). [Online]. http://data.worldbank.org/indicator/EG.ELC.ACCS.UR.ZS?view=maps. Accessed October 2023
4. Díaz, M., et al.: Edition September 12, 2022 Agency for the Regulation and Control of Energy and Non-Renewable Natural Resources (2022)
5. Ayala-Pico, J., Arcos-Aviles, D., Ibarra, A., Fernandez, C., Guinjoan, F., Martinez, W.: Current development of electricity generation systems in the Galapagos Islands – Ecuador. Renew. Energy Focus. **46**, 88–102 (2023). https://doi.org/10.1016/j.ref.2023.06.003
6. Xu, T., Gao, W., Qian, F., Li, Y.: The implementation limitation of variable renewable energies and their impacts on the public power grid. Energy. **239**, 121992 (2022). https://doi.org/10.1016/j.energy.2021.121992
7. Tapia, M., Heinemann, D., Ballari, D., Zondervan, E.: Spatio-temporal characterization of long-term solar resource using spatial functional data analysis: understanding the variability

and complementarity of global horizontal irradiance in Ecuador. Renew. Energy. **189**, 1176–1193 (2022). https://doi.org/10.1016/j.renene.2022.03.049
8. Esmaeel Nezhad, A., Mobtahej, M., Javadi, M.S., Nardelli, P.H.J., Sahoo, S.: Optimal operation of lithium-ion batteries in microgrids using a semidefinite thermal model. Int. J. Electr. Power Energy Syst. **155**(PB), 109630 (2024). https://doi.org/10.1016/j.ijepes.2023.109630
9. Benlahbib, B., et al.: Experimental investigation of power management and control of a PV/wind/fuel cell/battery hybrid energy system microgrid. Int. J. Hydrog. Energy. **45**(53), 29110–29122 (2020). https://doi.org/10.1016/j.ijhydene.2020.07.251
10. El-Faouri, F.S., Alzahlan, M.W., Batarseh, M.G., Mohammad, A., Za'ter, M.E.: Modeling of a microgrid's power generation cost function in real-time operation for a highly fluctuating load. Simul. Model. Pract. Theory. **94**(January), 118–133 (2019). https://doi.org/10.1016/j.simpat.2019.01.002
11. Arafat, M.Y., Hossain, M.J., Alam, M.M.: Machine learning scopes on microgrid predictive maintenance: potential frameworks, challenges, and prospects. Renew. Sustain. Energy Rev. **190**(PA), 114088 (2024). https://doi.org/10.1016/j.rser.2023.114088
12. Kim, B., Kim, I.: A case study of stand-alone hybrid power systems for a data center using HOMER and DIgSILENT. Energy Rep. **9**, 1136–1143 (2023). https://doi.org/10.1016/j.egyr.2023.02.044

Indoor Solar Thermal Cook Stove Design

Nelushi Munasinghe, Chansi Mendis, Nalika Ilukpitiya, and Inoka Manthilake

1 Introduction

Incorporating renewable energy sources, such as solar energy, which is a clean energy, into domestic cooking practices offers a transformative approach toward achieving energy independence from foreign sources. Especially, for developing countries such as Sri Lanka which lack foreign exchange reserves for importing fuel and, with world heritage natural habitats of highest biodiversity that get adversely affected from fossil fuel emissions. This method, which primarily leverages the sun's abundant energy for cooking, presents an environmentally friendly, zero emission, and sustainable alternative, for regions where solar exposure is richly available. The adoption of solar energy for cooking plays a crucial role in reducing reliance on fossil fuels, thereby contributing to a decrease in the carbon footprint and promoting a more sustainable future. The economic feasibility of solar cooking systems is a critical aspect, making them a viable option for many households. While the initial investment in a solar cooker may be higher compared to that of a conventional cooking burner, the operational costs associated with solar cookers are almost negligible. Such a system not only offers potential savings in overall energy consumption but also significantly reduces utility costs for households. This paper proposes a transformative approach to domestic cooking, aligning with global efforts to adopt renewable energy sources and reduce environmental impact.

In order to get a clear understanding of solar energy for cooking, a literature review was conducted under three topics, which were domestic cooking, availability of solar energy, and harvesting solar energy for cooking.

N. Munasinghe (✉) · C. Mendis · N. Ilukpitiya · I. Manthilake
Department of Mechanical Engineering, University of Moratuwa, Moratuwa, Sri Lanka

L. Chen (ed.), *Trends in Clean Energy Research*, Green Energy and Technology,
https://doi.org/10.1007/978-3-031-67987-2_4

1.1 Domestic Cooking

Wood, charcoal, kerosene, and liquefied petroleum gas (LPG) are generally the primary cooking fuels in Sri Lanka. Since the economic crisis has led to inflation and dismissal which affected the incomes of people in Sri Lanka, the cooking fuels became less affordable and less accessible for low-income households. On the other hand, using these traditional cooking fuels resulted in a significant environmental damage. However, including Sri Lanka, around three billion people (40% of the world's population) lack access to clean cooking fuel, while an estimated 940 million people (or 13% of the population) lack access to electricity which are the most environmentally friendly cooking fuel alternatives [1].

1.2 Availability of Solar Energy

The highest solar insolation in the world is found near the equator, particularly around the Tropic of Cancer and the Tropic of Capricorn, due to the direct and constant exposure to the sun's rays. The average solar insolation for these regions is around 6–7 kWh/m^2 per day. As we move away from the equator, solar insolation decreases. The average solar insolation for middle latitudes is around 4–5 kWh/m^2 per day, while at high latitudes, such as in northern Canada or Siberia, the average solar insolation is around 2–3 kWh/m^2 per day [2]. In Sri Lanka, solar insolation is generally high due to the country's proximity to the equator. The average solar insolation in Sri Lanka is about 4–6 kWh/m^2 per day, with the highest levels occurring during the months of March to September [Fig. 1a]. Hence Sri Lanka is enriched with solar energy which can be used in cooking scenarios as the energy source. However, it is important to note that the actual solar insolation will vary depending on factors such as atmospheric conditions and cloud cover. The southwest monsoon (May–September) brings in more clouds and rainfall to the southwestern and western parts of the country, and the northeast monsoon (December–February) brings more clouds and rainfall to the northeastern and eastern parts of the country [Fig. 1b] [3].

1.3 Solar Energy for Cooking

Solar energy is used as a direct and indirect form of solar radiation energy for cooking applications. Using solar thermal energy by concentrating solar radiation with the help of a collector or transmitting that energy through a fluid is categorized in the direct form of solar energy. Converting solar radiation energy to electricity is categorized as the indirect form of solar energy. In the indirect form, there are several energy conversions in solar electric hot plates and solar induction cookers,

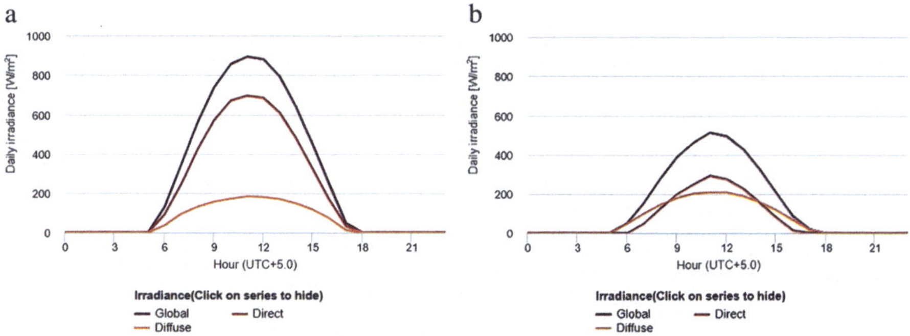

Fig. 1 (**a**) Daily average variation of solar irradiance in March for Sri Lanka. (**b**) Daily average variation of solar irradiance in June for Sri Lanka

even though the cooker unit has a significantly high efficiency [4, 5]. The overall efficiency is comparatively low due to the several terms of power losses [6]. In solar thermal-powered applications, solar thermal power plants such as solar parabolic troughs, solar parabolic dishes, and solar towers are examples of large-scale applications and, for domestic applications, solar water heating systems and solar cookers [7].

Existing solar cook stoves have explored various designs, including box cookers, panel cookers, parabolic cookers, and tube cookers [8]. Though these cookers have proven effective in harnessing direct solar radiation to achieve considerable cooking temperatures, they are unable to support indoor cooking and during periods of limited or no sunlight, such as nights and rainy days. These limitations have been a focal point in previous research, as they pose challenges for users who may prefer or necessitate indoor meal preparation day and night. Therefore, the investment made on those solar cook stoves does not pay back effectively. Hence, these are critical concerns in the design and development of solar cook stoves.

Having several energy conversions causes energy losses and reduces the overall efficiency. Hence, it is necessary to have minimum energy conversions from harnessing solar radiation to supplying heat at the stove. The output temperature has to be higher than 100 °C, in order to boil water. Through a solar collector, a fluid that can store thermal energy can be used to take up the sun's energy and transport it to a heat exchanger, where the heat is supplied to a cook stove. A fluid having high thermal stability, high thermal conductivity, low viscosity, low freezing point, good chemical compatibility, non-toxicity, and biodegradability such as mineral oil or synthetic esters was selected as the thermic oil [9].

The following research gaps could be identified considering the existing solar cook stoves.

- In indirect solar cookers, the overall system efficiency is low due to having many energy conversions.
- Having to be under the sun when using conventional solar cookers.
- Not being able to use during the nighttime and cloudy days.

- Not being able to produce temperatures far beyond 100°C.

A solar cooker was designed, fabricated, and tested to address the above research.

2 Methodology

This project followed three phases for the data collection and analysis. At first, concept designing, design calculation, and simulation were done. Then fabrication and prototype manufacturing and finally testing and validation were done to reach the aim of the project.

2.1 *Concept Designing*

The components of the proposed system are a solar energy collector, two thermic oil storages, a cooking stove, a thermic oil pump, pipelines, and a control panel. The system utilizes thermic oil as the heat transfer medium. The schematic diagram of the system is shown in Fig. 2.

The thermic oil is heated in the solar collector and stored in the first thermic oil storage tank. The cooking stove is equipped with a cooking hob designed to allow the flow of heated oil, creating a uniform heat distribution and flow behavior for efficient cooking. After transferring heat to the cooking pot, the outlet oil is circulated back to the second thermic oil storage tank.

The outlet from the low-temperature oil tank is subjected to the collector when there is adequate irradiation is present. Controlling the temperature is possible in this cook stove by controlling the fluid flow rate by changing the rpm of the motor. As

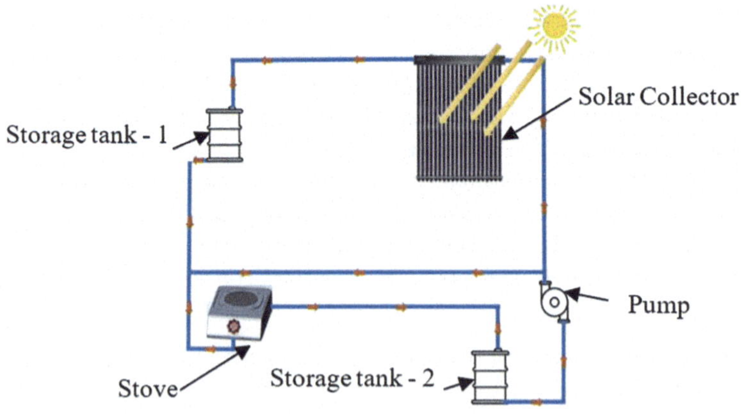

Fig. 2 System design

an alternative for solar energy, an electrical heater in the second thermic oil storage tank is proposed to be used in conditions where there is not an adequate amount of solar energy present more than the rated period. The system includes temperature control capabilities, allowing for adjustments based on cooking requirements.

2.2 Design Calculations

Energy Requirement for Food Consumption Assume that an average family with four family members required 2 kg of rice per day after conducting interviews with a few general families. The estimated energy consumption for cooking per day of a middle-income family in Sri Lanka is 17,370 kJ as shown in Table 1.

Water Boiling Apart from cooking, households use cooking fuel for boiling water several times per day. Therefore, boiling water was taken as the performance evaluating task, and to take the average energy consumption per day in an average family, a survey was conducted. There were 141 people who participated in the questionnaire survey from different provinces of the country. The data gathered from the survey is given in Table 2, and the average water boiling time was 5 min with the existing cook stoves.

Cook Stove Energy Calculation The energy required for water boiling per time using gathered data from the survey is 315 kJ using Eq. (1), where Q is the required energy, m is the mass of the water (1 liter) used per time, and $\Delta\theta$ is the maximum temperature difference (100–25°C) of the water boiling. Since the expected time to

Table 1 Power consumption for cooking meal

For average meal	Per meal		Per day (× 3 meals)	
Power (of the LPG cooking stove)	2050	W		
Time-consuming	40	min		
Energy consumption for cooking four curries	4,920,000	J	14,760,000	J
Power (of the rice cooker)			700	W
Time-consuming			30	min
Energy consumption for cooking rice			1,260,000	J
Power (of the electric kettle)	1500	W		
Time-consuming	5	min		
Energy consumption for water boiling	450,000	J	1,350,000	J
Total energy consumption per day			17,370	kJ

Table 2 Data gathered from the questionnaire survey

Requirements		Units
Amount of liter boil at a time	1	1
Water boiling time	5	Minutes
Water boiling times per day	3	

boil water from the survey is 5 min, the required heat supply rate for water boiling is 1.05 kW:

$$Q = mc\Delta\theta \tag{1}$$

In the proposed solar cooker, that amount of heat is supplied using a thermic oil. It is assumed that the oil temperature is maintained at 150 °C. Energy absorbed by the stove which is made by aluminum is 65 kJ by substituting to Eq. (1) the same as above, and to calculate heat losses, convection and heat conduction through walls were considered:

$$E_{\text{loss}} = \Delta T\left(\frac{1}{hA} + \frac{1}{kA}\right) \tag{2}$$

The total heat rate required for the cook stove is the summation of heat losses and energy absorbed by the water. To identify the minimum oil flow rate required for this task, Eq. (3) was used where the $\dot{m}$ is the flow rate of the oil and $\dot{Q}$ is the minimum heat rate required for water boiling:

$$\dot{Q} = \dot{m}C\Delta\theta \tag{3}$$

Solar Collector Sizing The average solar irradiation of Sri Lanka is 5.5kWh/m^2 per day. And the minimum requirement is to heat up the oil to 150 °C by using solar-evacuated tubes to absorb the solar irradiation energy. The chosen standard evacuated tube is with area of 2.5 m^2 and the efficiency is 78% [10]. Energy absorbed by the evacuated tube is 1014 W according to the solar irradiation of Sri Lanka, and throughout the calculations, the suggested amount of evacuated tubes is five with a length of 2 m.

2.3 Design and Simulations

The user requirements considered for designing the cook stove are the size of the stove which can fit into the kitchen, safety, cleanliness, and user friendliness. For sizing, an average domestic kettle with a bottom diameter 180 mm was considered. For effective heat transfer and minimum heat loss by calculations and simulations, the stove plate is designed with 166 mm diameter. Three final designs (Design A, spiral coil; Design B, maze pattern I; Design C, maze pattern II), as shown in Fig. 3, were investigated for the effective flow of the heat transfer fluid in the stove. The spiral coil is designed similar to an existing hot plate coil, and the rest of the designs are modeled using the above dimensions and based on maze patterns. The thickness of the inner plates is 3 mm and the outer wall thickness is 5 mm. For the material of the stove, plate aluminum, copper, mild steel, and cast iron were compared, and according to the comparison, aluminum was selected for the plate material since it

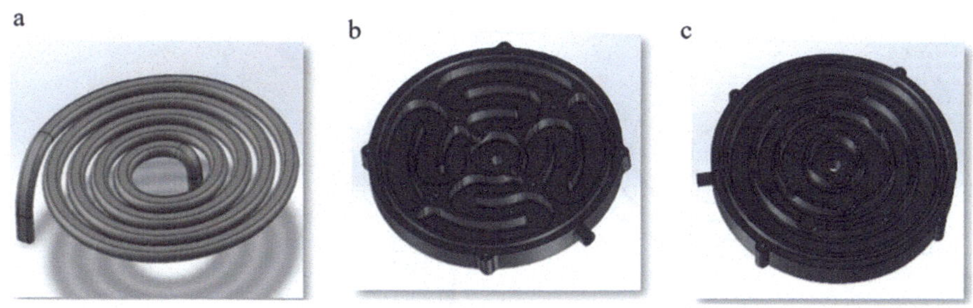

Fig. 3 (**a**) Design A, spiral coil; (**b**) Design B, maze pattern I; and (**c**) Design C, maze pattern II

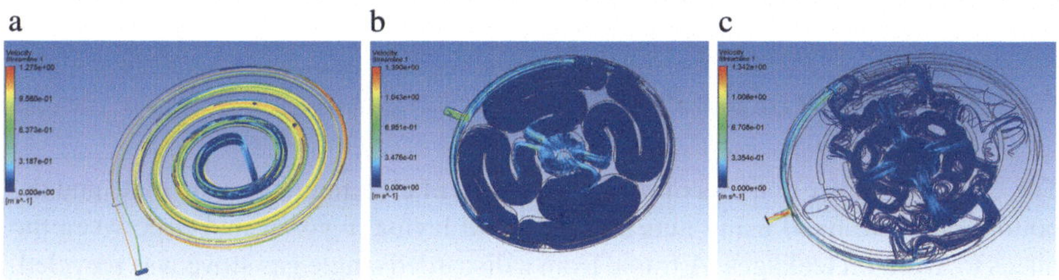

Fig. 4 (**a**) Fluid flow behavior of the spiral coil design, (**b**) fluid flow behavior of maze pattern I, and (**c**) fluid flow behavior of maze pattern II

has high thermal conductivity and low heat capacity, its cost is reasonable, and its manufacturability is also easy.

In all three designs, the inlet was designed in the center, and the outlet was designed to the outer edge of the stove. For the purpose of comparing flow behavior and temperature distribution, CFD simulations were carried out, and the most suitable design was selected for the cook stove. The simulations were performed on Ansys Workbench 19.2 which is a simulation integration platform using the tools of fluid flow (fluent).

As can be seen in Fig. 4, the A and B designs have smooth flow patterns, while design C has stagnant and imbalanced flow behavior. Design C, maze pattern II, further leads to low-temperature surface area of the top plate of the stove. Then the simulation results of the temperature distribution for the A and B designs were compared.

The designs were constructed to obtain a similar temperature behavior to existing cooking stoves. In comparison with the temperature behavior of the existing cookers [11, 12], even Fig. 5b simulation represents a significant similarity in CFD simulation outputs. Therefore, the Design B cook stove represents more similarity with the existing cook stoves. Hence, for comparison of both flow rate and temperature behaviors, "Design B" was selected as the most suitable design for fabricating the cook stove.

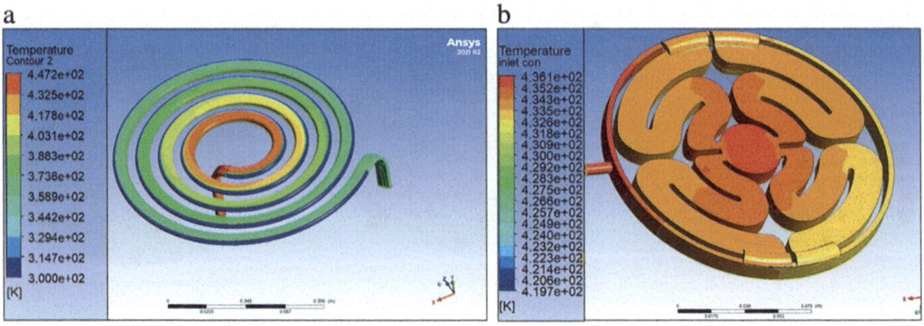

Fig. 5 (**a**) Temperature distribution of Design A and (**b**) temperature distribution of Design B

2.4 Fabrication of the Prototype

The cook stove plate consists of two plates made of aluminum. The top plate and the bottom plate are fixed using nuts and bolts and laying a gasket maker between the plates to avoid oil leakages. A frame from a discarded single gas stove was recycled, and the stove plate was settled in it with the surrounding layer of clay insulation. A metal tank was used as the storage tank which can withstand high-temperature ranges. Since a solar-evacuated tube collector cannot be adopted in the prototype due the high cost for the process of stage 1, a heating coil of the same wattage of 3000 W was submerged in the thermic oil in high-temperature storage tank. The heating coil was attached through a thermostat switch. The inlet to the tank comes from the oil pump, and the outlet of the tank was subjected to the stove plate.

A control system was developed for the cook stove, and it consists of an indicator that displays the top plate temperature, cooking power level (low, medium, high), storage tank temperature, the remaining cooking time in minutes to use existing oil in the storage tank, and oil level. Also, the system indicates the need for pump to switch on or off, with a color-changing LED. When the cook stove is on, it is indicated by a green light. Figure 6 shows the stove plates and the control system-integrated cook stove.

2.5 Prototype Functionality Testing

First, the thermic oil flow behavior was tested by using a colored fluid that was subjected through the cooking stove plate, which is shown in Fig. 7a. Then the control system was tested to get the required readings and data to operate at the required procedures, which are shown in Fig. 7b, c. With that, the flow rate sensor readings are used to evaluate whether the required fluid flow rate could be achieved or not. The heating coil that is implemented inside the tank was powered up, and the oil is heated to a maximum of 120 °C under the proper switching of the thermostat.

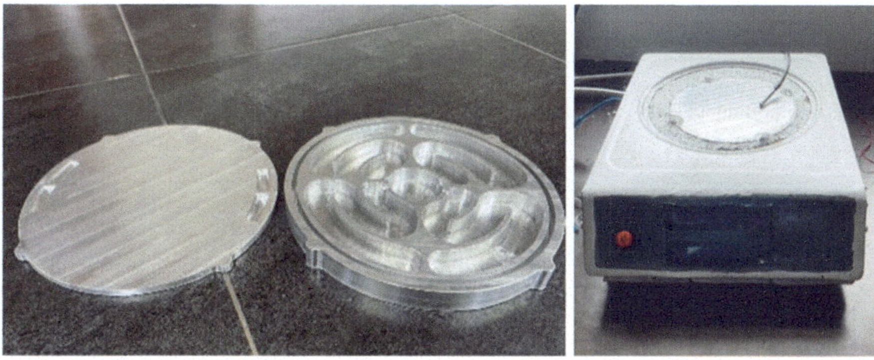

Fig. 6 Prototype of the solar cook stove

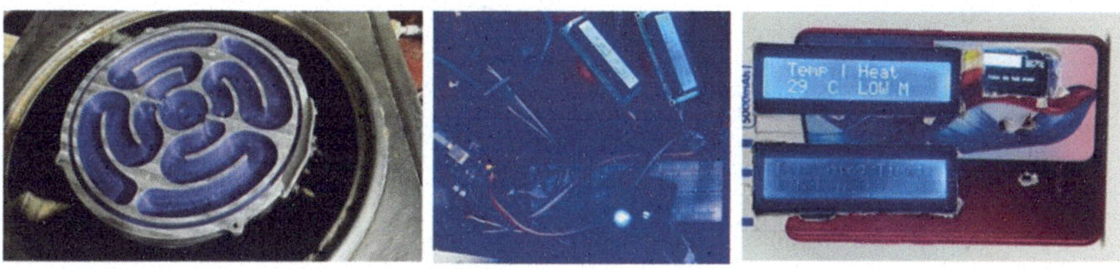

Fig. 7 (**a**) Functionality testing by using a colored fluid, (**b**) functionality testing of the control system, and (**c**) functionality testing of the displays

Then the heated oil was subjected through the stove plate to evaluate the time taken for 1 liter of water (at room temperature) to get boiled.

3 Experimental Results and Discussions

As for the prototype, alternatives for the thermic oil and solar collector were used at the current stage, and testing was done according to them. The following test results were taken by using coconut oil as a thermic oil and a heating coil as solar-evacuated tubes. The temperature of the top plate surface, oil inlet and outlet temperature, and the temperature of water were read using a temperature sensor. The data were taken by adjusting the thermostat to a temperature of 120 °C and evaluating the time taken for 1 liter of water (at room temperature) to get boiled. The temperature variation of water, stove plate, and oil outlet when the coconut oil is supplied to the stove plate at 120°C is shown in Table 3 and Fig. 8.

It can be observed that the time taken to increase the temperature by the same increment (20 °C) increases as the temperature of the water rises. The temperature difference reduction with time is the reason for the time taken to increase the temperature of water by 20 °C. Therefore, it can be deducted that if the temperature

Table 3 Temperature variation of water, stove plate, and oil outlet when the coconut oil is supplied to the stove plate at 120 °C

Flow rate (cm^3/s)	Temperature of the oil inlet (T°C)	Temperature of the water (T°C)	Time taken to heat up water (mins)	Time taken to heat up water (seconds)
22.22	120	25	0	0
		40	5	281
		60	7	445
		80	11	685
		100	15	880

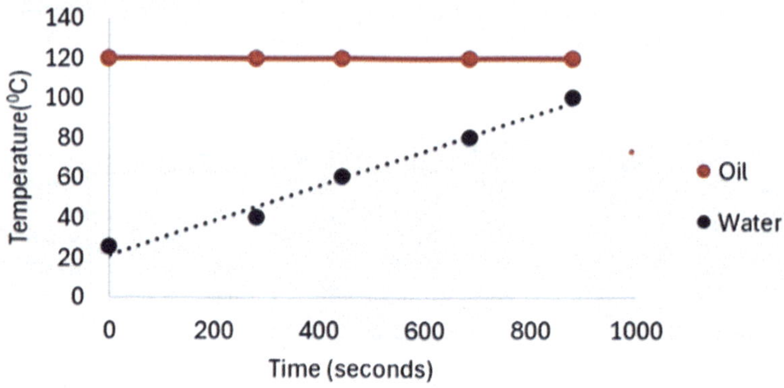

Fig. 8 Temperature variation of water when the coconut oil is supplied to the stove plate at 120 °C

of the oil inlet exceeds 120 °C, the boiling time for water is shorter than 880 s (less than 15 min). In the prototype, the oil inlet temperature was not allowed to rise due to safety concerns. The prototype is being further improved with the solar-evacuated tubes instead of the heating coil and proper insulations, to reach the expected efficiency as in the simulations.

4 Conclusions

In contrast to many existing solar cookers, the proposed system focuses on achieving the requirement of "indoor cooking," without direct exposure to sunlight and the risk of sunburn. Utilizing solar thermal power ensures better efficiency, compared to using electric cookers that run on solar photovoltaic-generated power, since having several energy conversions increases opportunities of energy loss, thus reducing efficiency. Additionally, the system capitalizes on cooking capabilities through its high-temperature output, exceeding 100 °C.

The designed solar-powered cook stove system with energy storage demonstrated capabilities comparable to a conventional electric hot plate and provides clean cooking opportunities over the existing LPG cookers. As the solar stove is equipped with energy storage, ensuring that cooking times are not limited to daytime. Additionally, the stove can be utilized with off-peak electricity, thus helping to reduce peak loads on the electricity grid in areas where peak demands are driven by domestic usage by using a heating coil. The most highlighted limitation is lengthy periods of rainy days, and to overcome such conditions, an alternating method, such as electric heating, can be implemented to heat up the oil in storage tank-II. The storage tank is able to supply a thermal energy for 6 h with the condition of consuming average energy consumption per meal which is sufficient for daily cooking. Therefore, the proposed cook stove system and design successfully address the identified research gap of a solar cook stove by achieving the mentioned key requirements altogether. The external battery power to supply electrical energy for the control system and to run the motor of the pump to circulate the oil is only 40 W. Further experimentations are being carried out with thermic oil and evacuated tube solar collectors for reducing the initial cost for solar collectors and piping networks to implement this system as a domestic cooker affordable to a larger population.

Acknowledgements The authors gratefully acknowledge the invaluable support and unwavering guidance from the families of the authors and the academic, nonacademic, and workshop staff of the University of Moratuwa. Additionally, the authors would like to extend special thanks to Romasha Engineers Pvt. Ltd. and the University Business Linkage Cell (UBLC) of the University of Moratuwa for their support in advancing the project further.

Declaration of Competing Interest The authors declare that they have no known competing financial interests or personal relationships that could have appeared to influence the work reported in this paper.

Data Availability Data will be made available on request.

References

1. Jayasinghe, M., Selvanathan, S., Selvanathan, S.: Energy poverty in Sri Lanka. Energy Econ. **101**, 105450 (2021). https://doi.org/10.1016/j.eneco.2021.105450
2. Solar Resource Maps and data: Geospatial Data Science | NREL (n.d.). https://www.nrel.gov/gis/solar-resource-maps.html
3. User S. Climate of Sri Lanka (2019). https://www.meteo.gov.lk/index.php?option=com_content&view=article&id=94&Itemid=310&lang=en
4. Batchelor, S., Talukder, M.Z., Uddin, M.K., Mondal, S., Islam, S., Redoy, R., et al.: Solar e-cooking: a proposition for solar home system integrated clean cooking. Energies. **11**, 2933 (2018). https://doi.org/10.3390/en11112933
5. Anusree, K.V., Sukesh, A.: Solar induction cooker. In: IEEE Conference Publication | IEEE Xplore 2020. https://ieeexplore.ieee.org/document/9339725
6. Photovoltaic Energy Factsheet: Center for Sustainable Systems (n.d.). https://css.umich.edu/publications/factsheets/energy/photovoltaic-energy-factsheet
7. Solar thermal power plant – Energy Education (n.d.) https://energyeducation.ca/encyclopedia/Solar_thermal_power_plant
8. Shukla, K.: An overview study of solar cookers. Parul (2019)
9. Heat transfer fluids for solar water heating systems. (n.d.). Energy.gov. https://www.energy.gov/energysaver/heat-transfer-fluids-solar-water-heating-systems

10. Govindasamy, K., Palanivel, V., Meena, R.S., Muthusamy, S., Panchal, H., Shah, M.A., et al.: Performance analysis of evacuated tubes with thermosyphon heat pipe solar collector integrated with compound parabolic concentrator under different operating conditions. Energy Explor. Exploit. (2023). https://doi.org/10.1177/01445987231202618
11. Matthujak, A., Wichangarm, M., Sriveerakul, T., Sucharitpwatskul, S., Phongthanapanich, S.: Numerical investigation on the influences of swirling flow to thermal efficiency enhancement of an LPG-energy saving burner. Case Stud. Therm. Eng. **28**, 101466 (2021). https://doi.org/10.1016/j.csite.2021.101466
12. Chen, W.L., Liao, K.-C., Wu, H.-Y., Wen, H.-T.: Heat flow and thermal stress analysis to enhance the temperature distribution and service life of stone grill plates in barbecue ovens. Inventions. **7**, 103 (2022). https://doi.org/10.3390/inventions7040103

Solar Energy Performance Prediction with Regression Algorithm in Machine Learning Based on Weather Conditions: A Case Study

Atıl Emre Coşgun

1 Introduction

The world's energy consumption, especially for electrical energy, is continuously increasing due to population expansion, contemporary lifestyles, and increased technology. All over the world, fossil fuel energy sources play a major role in supplying energy requirements. However, it not only meets our energy needs but also causes global warming. Therefore, presently, energy technologies are transitioning toward the utilization of renewable energy sources to mitigate the effects of global warming. Solar photovoltaic (PV) energy systems are the most preferred options for renewable energy. Several nations are transitioning to the utilization of solar photovoltaic (PV) energy-generating systems to produce electricity on a large scale. The reason for this is that it is an environmentally friendly and sustainable energy source, requiring minimal expenses for upkeep [1]. However, a significant challenge associated with renewable energy sources is their inherent unpredictability. Photovoltaic (PV) plants produce varying amounts of power depending on the fluctuations in solar radiation and other meteorological factors [2]. A primary obstacle to the use of solar PV power is its reliance on sunshine, which varies owing to factors such as weather patterns, time of day, and seasonal fluctuations. Producing electric energy might vary significantly not just across various days but also from 1 h to another, or even within a few minutes [3]. In this situation, the presence of PV power on the grid can result in a mismatch between power demand and supply. Therefore, precisely predicting solar PV power generation is crucial to optimizing its integration into the grid, maintaining stability, and maximizing its potential advantages. This is especially critical in networks with a high level of PV integration. Also, estimating the power output of a photovoltaic (PV) system is crucial for many reasons.

A. E. Coşgun (✉)
Department of Software Engineering, Aksaray University, Aksaray, Turkey

L. Chen (ed.), *Trends in Clean Energy Research*, Green Energy and Technology,
https://doi.org/10.1007/978-3-031-67987-2_5

Initially, precise photovoltaic (PV) power estimations guarantee that a solar system is appropriately dimensioned to fulfill consumers' energy requirements. In addition it effectively regulates the balance between energy supply and demand, therefore guaranteeing sufficient power availability to fulfill the requirements of users. PV power estimations are furthermore employed to ascertain the economic feasibility of a solar endeavor. Precise assessments of the power generation of a photovoltaic (PV) system enable investors, lenders, and other financial players to make well-informed choices on their investments in a specific solar project [4]. Furthermore, accurately determining the power generated by a photovoltaic (PV) system is essential in terms of effectively monitoring its performance and doing necessary repairs too. Through the comparison of projected energy generation with real measured values, individuals who own and operate solar photovoltaic (PV) systems can evaluate the system's efficiency, detect any deviations or subpar performance, diagnose potential problems, and implement necessary measures to guarantee optimal functioning and maximize energy output throughout the system's lifespan. Additionally, the diversity of locations in exploiting the solar resource adds additional complexity in estimating energy production. The variability of solar irradiance is influenced by factors such as latitude, altitude, cloud cover, and atmospheric conditions, which poses difficulties in creating universally applicable forecasting models. Places located closer to the equator may have a more uniform amount of sunshine throughout the year in comparison to places at higher latitudes, where there are more noticeable changes in sunlight due to the seasons. Therefore, in order to offer accurate forecasts customized to certain areas, forecasting algorithms must take into consideration these regional disparities. There are many studies in the literature to overcome these challenges. Generally, they have benefited from some technologies, such as providing real-time data from weather satellites, ground-based sensors, and historical weather trends. Thus, by trying different methods, researchers have anticipated variations in solar PV power output through satellite imaging, weather modeling, machine learning algorithms, and advanced data analytics. According to Gupta and Singh [5], solar power forecasting may be categorized into two groups: (1) direct forecasting, which predicts PV power directly, (2) indirect forecasting, which relies on the prediction of solar radiation. However, there are studies evaluating short-term solar power prediction under four different study headings [6, 7]. These are physical, statistical, machine learning, and deep learning. Physical approaches utilize physical data, such as the amount of sunshine received by solar panels, to make forecasts. Physical approaches do not need previous solar power data, but they are extremely responsive to initial parameter configurations (such as performance parameters and panel tilt angle). Consequently, they are frequently integrated with other methods to enhance precision [8, 9]. Solar energy production forecasting is the estimation of the quantity of solar energy that will be created within a defined timeframe, usually spanning from hours to days in advance. Statistical approaches are crucial in this process as they allow for precise forecasts by analyzing historical data and considering numerous affecting factors. Machine learning provides a versatile and data-based method for predicting solar energy output. It achieves accurate forecasts by utilizing historical data, pertinent characteristics, and sophisticated modeling approaches. Also, using this strategy is an efficient way to discover nonlinear connections between several variables and the aim for forecasting. Some often used approaches in machine learning are regression models (RM), principal component analysis (PCA), weighted generalized partial residuals (WGPR), support vector machines (SVM), random forest (RF), gradient boosting decision trees (GBDT), and expectation maximization learning (EML). Deep learning encompasses the utilization of deep neural network models, which include the ability to autonomously extract data and make predictions. The adaptable characteristics of network architectures have resulted in the creation of several deep learning models, including artificial neural networks (ANN) and long short-term memory (LSTM). According to the existing research, using machine learning is an effective method for forecasting solar radiation. With this understanding, we made the decision to utilize to predict the quantity of energy produced by solar power plants depending on various weather-based conditions in the Konya region. The reason for the selected city is that the solar power plants installed in Turkey have the largest share in terms of production values and energy production potential. This paper presents an examination of regression algorithm forecasting approaches in machine learning for predicting the quantity of energy produced by solar power plants with various weather conditions. Regression algorithms utilize training data to construct a mathematical model capable of forecasting the output of a photovoltaic (PV) system, taking into account input factors such as solar radiation and temperature. Regression algorithms provide the capability to forecast both short-term and long-term photovoltaic (PV) power outputs [10]. The research questions we will answer are as follows: (1) Which forecasting approach is optimal for specific places with minimal data? (2) What impact does it have on the value of solar energy production in relation to various weather conditions? (3) What are the ramifications of our discoveries for predicting, formulating policies, and establishing regulations? These questions can also be used in many regions with minimal research or limited availability of data. This study was conducted in Karapınar, one of the largest towns of Konya (2623 km^2), which is specifically located in the east of the province (37 °0.71′N, 33° 54′ E). The data was obtained from an unlicensed 500kWe solar energy plant. The electricity generation data for the year 2023 has been collected at hourly, daily, and monthly intervals. The location chosen is based on the fact that Turkey's solar power plants have the highest proportion of production values and energy production potential. Furthermore, meteorological data specific to the local area is accessible for the duration of the research term.

2 Material and Methods

In this study, the characteristics taken into account for meteorological data are monthly total precipitation average (mm = kg/m^2), average sunshine duration (hours), monthly average temperature (°C), monthly average number of rainy days (mm = kg/m^2), monthly average wind speed (m/s), monthly average 5 cm. soil temperature (°C), monthly solar irradiation estimates (kWh/m^2), monthly average relative humidity, and aerosol optical depth, 550 nm. The annual meteorological data of the relevant region were obtained from the Turkish State Meteorological Service. The monthly irradiance information was obtained from the European Commission's Photovoltaic Geographical Information System. Moreover, the aerosol optical depth data was obtained from NASA's Giovanni the Bridge Between Data and Science platform. The forecasting performance of a prediction model is directly influenced by the characteristics of the feature vectors utilized as input. In order to increase data consistency, normalization was performed on all data. Moreover, this process was carried out because data with different decimal values would prevent correct inference in our model. To realize this study, a support vector machine (SVM) model has been used. A SVM is considered as a supervised learning technique that is commonly employed for classification and regression tasks. It is particularly useful in many domains such as facial analysis and detection, handwriting recognition, pattern classification, and regression, among others. This model performs the prediction of the target variable by trying to find a hyperplane that provides the best fit between the samples in the data set. A classification approach involves the use of training and testing data properties. SVM has two key advantages: it reduces classification error through an iterative training procedure and maximizes the margin between two hyperplanes. In Fig. 1, the X axis is dependent on weather condition outputs for each month from January to September. Data was obtained from ground stations and satellites for that location. In the modeling carried out using MATLAB regression learner app, blue lines show the real values (true) and orange lines show the estimated values (predicted) produced by the model. As a result, it was observed that there was a correlation of 89% between the estimated values and the actual values according to a third-degree polynomial. The purpose of this is to train the

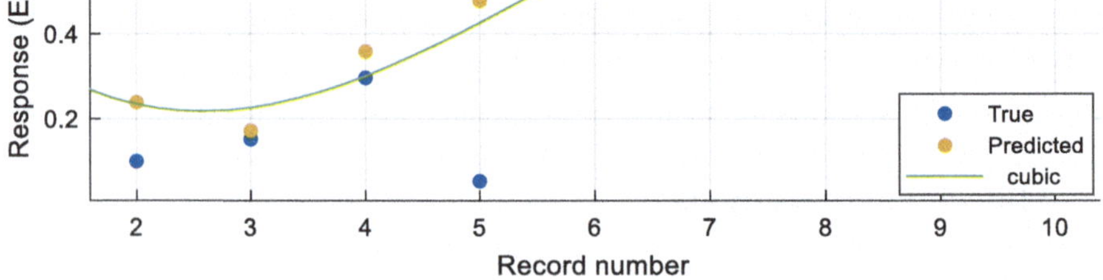

Fig. 1 Actual and estimated electricity production values

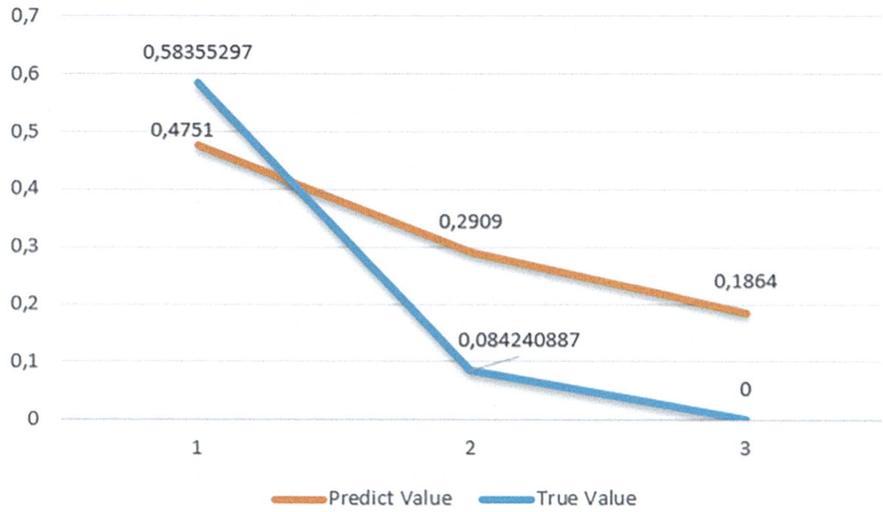

Fig. 2 Estimated production values based on ML Model 2

data based on the actual values and use the regression as a prediction for other values.

There are two lines in Fig. 2. One of them is true (known values), and the other is as a result of the predicted values. To do this, nine data from January to September are defined in the model. Three values from October to December were estimated as a result of inference according to the model relationship. Input data are mentioned before, and the output is the normalization value of the energy produced.

In addition, residual analyses were presented in Fig. 3. The residuals provide valuable information about the model's performance in relation to its real value. The graphic illustrates a linear regression model that exhibits asymmetrical scattering of about 0, substantial size variations from left to right, the presence of outliers, and a nonlinear pattern in the residual analysis across all techniques depicted.

Furthermore, in the study when different regression learning models were applied, the best result was obtained in the quadratic model (see Table 1). Moreover, for the performance evaluation of the model in terms of identifying the error percentage and for model suitability, regression metrics such as residual analysis, R-squared, root mean square error (RMSE), and mean absolute square (MAE) have benefited. Listed below are the equations and formulas for calculating errors in various approaches:

$$\text{RMSE} = \sqrt{\frac{1}{N}\sum_{i=1}^{N}\left(Y_{\text{pred}} - Y_{\text{true}}\right)^2} \tag{1}$$

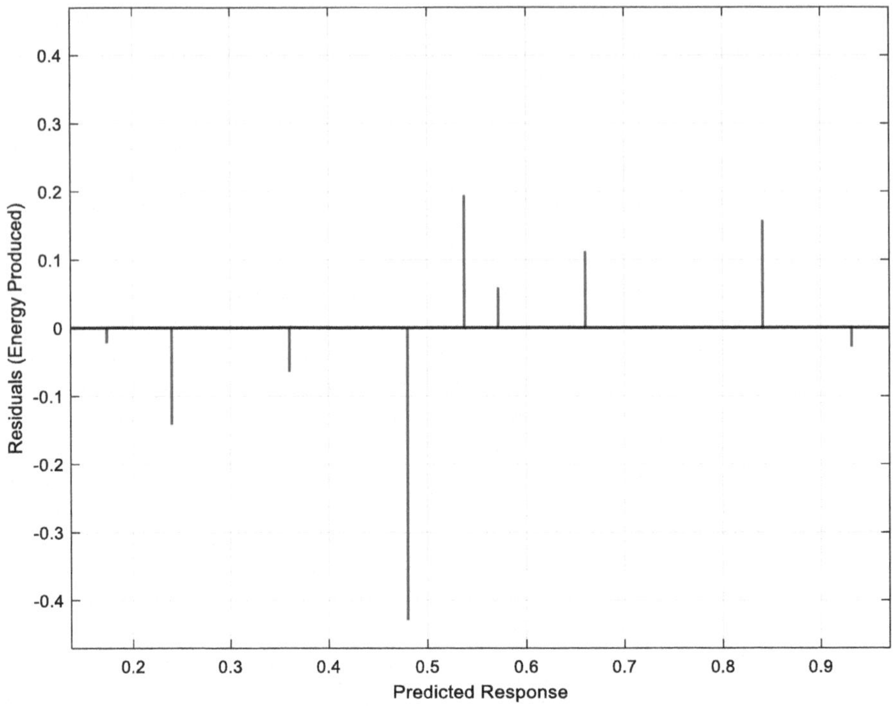

Fig. 3 Residual analyses

Table 1 Different regression learning model outcomes

Model	RMSE	R^2	MAE
1. Linear SVM	0.183	0.82	0.125
2. Quadratic SVM	0.178	0.83	0.134
3. Cubic SVM	0.237	0.69	0.189
4. Medium Gaussian SVM	0.224	0.72	0.183

Let N be the number of observations or iterations used to determine the error. Y_{pred} refers to the projected values of the dependent variable, whereas Y_{true} represents the true or actual value. True values refer to the input values provided to a trained model, whereas predicted values are the values retrieved by regression analysis.

The mean absolute error (MAE) is calculated by taking the absolute difference between the expected and real answers. The formula for MAE is shown below:

$$\text{MAE} = \frac{1}{N}\sum_{i=1}^{N}\left(Y_{\text{pred}} - Y_{\text{true}}\right) \tag{2}$$

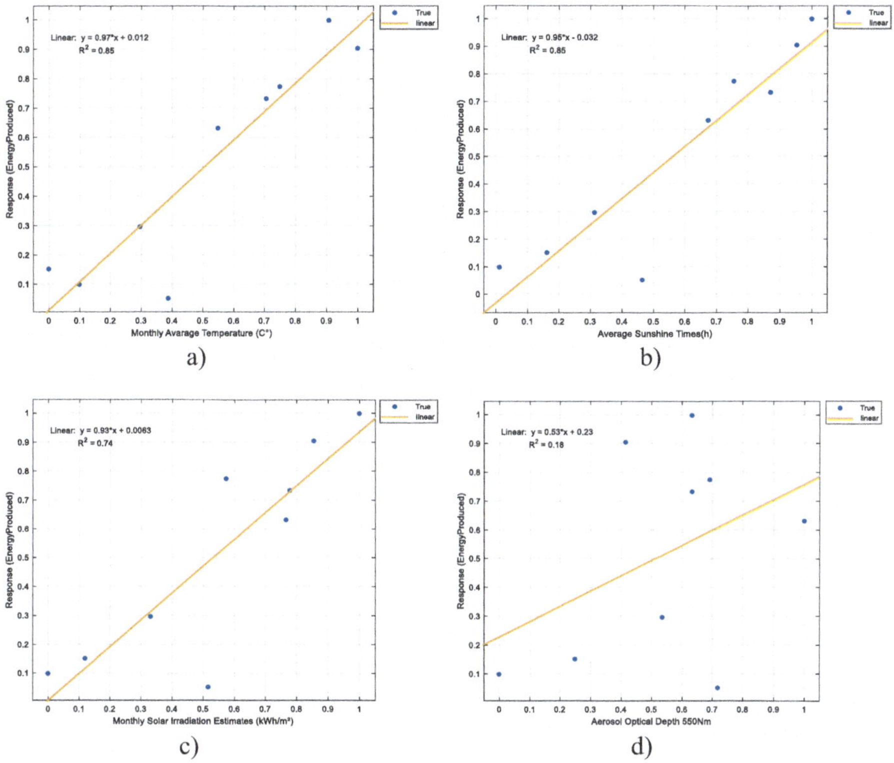

Fig. 4 Effect of different input values on output. (**a**) Monthly average temperature, (**b**) average sunshine times, (**c**) monthly solar irradiation, and (**d**) aerosol optical depth

The coefficient of R squared quantifies the amount of variation explained by the dependent variables. These errors cannot be measured on a scale and typically have a value of unity, regardless of the amount of the data. In study, the R^2 value was greatest for SVM regression Model 2, with a coefficient of 0.83. This indicates that 83% of the production of solar energy plants can be predicted by this model:

$$R^2 = 1 - \frac{\text{Residual Sum of Squares (RSS)}}{\text{Total Sum of Squares (TSS)}} \tag{3}$$

On the other hand, when the effects of data on energy production are examined one by one, it has been observed that temperature values have a greater effect (Fig. 4). The reason for the lower effect of solar irradiation on the solar panels could be that the panels are not clean. This is a variable that is not controlled in our study.

2.1 Comparison and Evaluation

Linear SVM and quadratic SVM exhibit almost similar R^2 values, with quadratic SVM slightly outperforming in both metrics. However, linear SVM has a lower MAE compared to quadratic SVM. Cubic SVM shows a higher RMSE and lower R^2 compared to linear and quadratic SVMs, indicating poorer performance in terms of accuracy and goodness of fit. Its MAE is also the highest among the listed models. Medium Gaussian SVM falls between the linear, quadratic, and cubic SVMs in terms of performance metrics. While its RMSE and R^2 values are better than cubic SVM, they are slightly inferior to linear and quadratic SVMs. However, its MAE is notably better than cubic SVM but worse than linear and quadratic SVMs. Overall, quadratic SVM seems to be the best-performing model among the listed options, considering its balanced performance across all metrics.

3 Conclusion

Harnessing solar energy effectively poses a multitude of challenges, chief among them being the intermittent nature of sunlight and its variability across different geographical locations and timeframes. As such, the need to forecast solar PV power production ensures stability and maximizes its potential benefits. As a result, there is a nonlinear relationship between input data and output responses. It is clear that depending on input values (the monthly total precipitation average (mm = kg/m^2), average sunshine duration (hours), monthly average temperature (°C), the monthly average number of rainy days (mm = kg/m^2), monthly average wind speed (m/s), monthly average 5 cm. soil temperature (°C), monthly solar irradiation estimates (kWh/m^2), monthly average relative humidity, and aerosol optical depth (550 nm), the solar plant performance could have relatable dependences at a rate of 0.83 values in terms of R^2 correlation. Among the input values, the most significant relationships that would affect the output performance were between average temperature, sunshine duration, and radiation values.

References

1. Iheanetu, K.J.: Solar photovoltaic power forecasting: a review. Sustain. For. **14**, 17005 (2022). https://doi.org/10.3390/su142417005
2. Ismail, A.M., Ramirez-Iniguez, R., Asif, M., Munir, A.B., Muhammad-Sukki, F.: Progress of solar photovoltaic in ASEAN countries: a review. Renew. Sust. Energ. Rev. **48**, 399–412 (2015). https://doi.org/10.1016/j.rser.2015.04.010
3. Kasu, M., Hara, S., Uematsu, T.: High-speed measurements of generated power and its relationship to weather observations at Yoshinogari Mega Solar Power Plant. In: Proceedings of the 2017 IEEE 44th Photovoltaic Specialist Conference (PVSC), pp. 1950–1953. IEEE, Washington, DC, June 2017

4. Sharadga, H., Hajimirza, S., Balog, R.S.: Time series forecasting of solar power generation for large-scale photovoltaic plants. Renew. Energy. **150**, 797–807 (2020). https://doi.org/10.1016/j.renene.2019.12.131
5. Gupta, P., Singh, R.: PV power forecasting based on data-driven models: a review. Int. J. Sustain. Eng. **14**, 1733–1755 (2021). https://doi.org/10.1080/19397038.2021.1986590
6. Guermoui, M., Melgani, F., Gairaa, K., Mekhalfi, M.L.: A comprehensive review of hybrid models for solar radiation forecasting. J. Clean. Prod. **258**, 120357 (2020). https://doi.org/10.1016/j.jclepro.2020.120357
7. Zhou, Y., Liu, Y., Wang, D., Liu, X., Wang, Y.: A review on global solar radiation prediction with machine learning models in a comprehensive perspective. Energy Convers. Manag. **235**, 113960 (2021). https://doi.org/10.1016/j.enconman.2021.113960
8. Park, B., Hur, J.: Spatial prediction of renewable energy resources for reinforcing and expanding power grids. Energy. **164**, 757–772 (2018). https://doi.org/10.1016/j.energy.2018.09.032
9. Zazoum, B.: Solar photovoltaic power prediction using different machine learning methods. Energy Rep. **8**, 19–25 (2022). https://doi.org/10.1016/j.egyr.2021.11.183
10. He, Y., Yan, Y., Xu, Q.: Wind and solar power probability density prediction via fuzzy information granulation and support vector quantile regression. Int. J. Electr. Power Energy Syst. **113**, 515–527 (2019). https://doi.org/10.1016/j.ijepes.2019.05.075

Part II
Renewable Energy Technology and Energy Efficiency Optimization

Technical-Economic Evaluation of a Stand-Alone Hybrid Renewable Energy System with Different Dispatch Strategies

Juan Lata-García, Pablo Parra, Gary Ampuño, and Lizbeth Flores-Bastidas

1 Introduction

Currently, around 2.4 billion people in the world do not have reliable access to electricity and other types of energy. The lack of electrification increases in isolated places in developing countries where the lack of electrical infrastructure due to cost or difficulty of implementation has not been realized [1]. The United Nations in 2015 adopted a series of strategies to eradicate poverty and care for and ensure the sustainability of the world. The 17 points to be worked on in the new agenda were called Sustainable Development Goals. Goal 7 refers to affordable and non-polluting energy, being the implementation of renewable systems a solution for electrification complying with the proposed agenda [2].

Academia and research centers have worked hard to improve the technical and economic aspects of the implementation of renewable energy systems, where the results demonstrate the viability of the use of hybrid systems. The term refers to when more than one of the clean energy sources is used, improving the reliability of the system [3–6]. However, there is still a need to deepen the research in the technical economic evaluation with different dispatch strategies.

Over time, researchers have joined forces to design and know the feasibility in the implementation of hybrid systems of isolated renewable energies. The main sources used are solar and wind. In the reference [7], Rezk and Dousoky supplied electrical energy produced by the system made up of solar panels, wind turbines, and fuel cells for remote areas. Six types of configurations are discussed: solar

J. Lata-García (✉) · P. Parra · L. Flores-Bastidas
Universidad Politécnica Salesiana – GIPI, Guayaquil, Ecuador
e-mail: jlatag@ups.edu.ec

G. Ampuño
Universidad Politécnica Salesiana – Smart tech, Guayaquil, Ecuador

L. Chen (ed.), *Trends in Clean Energy Research*, Green Energy and Technology,
https://doi.org/10.1007/978-3-031-67987-2_6

panels/batteries, solar panels/fuel cell, wind turbine/batteries, solar panels/wind turbine, wind turbine/fuel cell, and solar panels/fuel cell and batteries. In the document presented, the operation and performance for each of the configurations to be installed in Minya Governate (Egypt) is described in detail, to compare each of the systems and find the optimal configuration. The HOMER software is used. The computer program indicates the net current cost, the initial cost, and the levelized cost of energy. The configuration with the lowest cost and highest performance is the one made up of solar panels and fuel cell.

In a study conducted by Vaziri Rad et al. [8], they prioritized in an autonomous microgrid the economic part and reliability optimizing the high excess of electricity and the zero consumption of the load at certain times of the day. The management of the surplus of electricity after fully charging the storage subsystem imposed a negative impact on the levelized cost of energy. On the other hand, the reduction of excess electricity was taken into account when adjusting the power generation and the consumption profile of the load. In the research, the authors presented a novel dispatch strategy called modified predictive dispatch (MPD), which focuses on the proactivity of loading and unloading storage and consumption management. The results showed that excess energy can be reduced by 7.5%, compared to 41% and 18.5% for load tracking, while the rate of self-utilization of unused energy improved from less than 25–90%. The cost of energy also showed a reduction in value of 10% reaching $0.12/kWh. With the results presented, the predictive dispatch strategy is validated.

Ravi et al. studied the environmental and economic performance of a hybrid renewable energy system consisting of solar panels, wind turbine, diesel generator, and a battery bank using different energy dispatch strategies. The energy and economic results are compared to indicate the optimal configuration. The predictive energy strategy is carried out by HOMER for a 197.38 kW photovoltaic system, a 50 kW diesel generator, a battery bank of 599.76 kWh, and an inverter of 80.61 kW. The unsatisfied load corresponds to 0%, while the unsatisfied load is only 1.45 kWh/year. In the environmental issue, 93% of the use of renewable energy is observed with a 90% reduction in the emission of greenhouse gases [9].

2 Configuration Under Study

2.1 Location of Implementation

In the Gulf of Guayaquil, Ecuador, there are a number of communities that do not have access to conventional electricity. In the community of mass 2, there are 17 houses. The coordinates are 2°22.7′S and 79°51.6′W. The only way to generate electricity in the community is by diesel generator, which works for 3 h at night from 6:00 p.m. to 9:00 p.m. Each family pays 1 USD per day for the use due to the complexity of transporting fuel by sea. A survey of the loads used by each of the

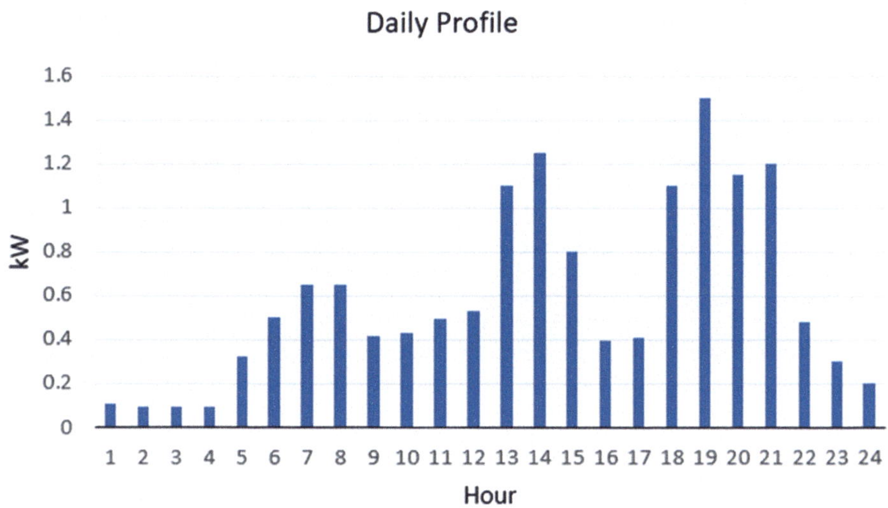

Fig. 1 Consumption profile of housing in the mass community 2

families was carried out, resulting in a maximum peak load of 1.5 kW. The daily energy consumed is 14.37 kWh/day. In Fig. 1, the load profile of the houses of the community is shown.

2.2 Solar Resource and System Component Modeling

The coastline where the community is located has abundant solar resources reaching an average of 4.58 kwh/m^2/day; however, the wind potential is quite low with a maximum of 3 m/s, so this type of technology is not viable to use. In Fig. 2, the solar potential is shown throughout the year.

The power generated by the photovoltaic system can be calculated using Eq. 1. Solar radiation plays the fundamental role. The variables of the equation are as follows: it is the power of the solar panel, it is the derating factor, it is the percentage of performance of the panel, and it is the irradiance on the panels measured in kW/$P_{PV}P_RD_{PV}S(t)^{\text{m2}}$, while it is the reference irradiance, it is the temperature coefficient, and it is the temperature of the solar cell, $S_{\text{ref}}\alpha_pT_{\text{cell}}$ [10]

$$P_{PV}(t) = P_RD_{PV}\left(\frac{S(t)}{S_{\text{ref}}}\right)\left[1 + \alpha_p\left(T_{\text{cell}}(t) - T_{\text{ref}}\right)\right] \quad (1)$$

The panel used in the work is from the manufacturer Trina monocrystalline type. The rated power is 345 Wp, the efficiency of the panel is 17.8%, and the cost per kW is 470 USD/kW.

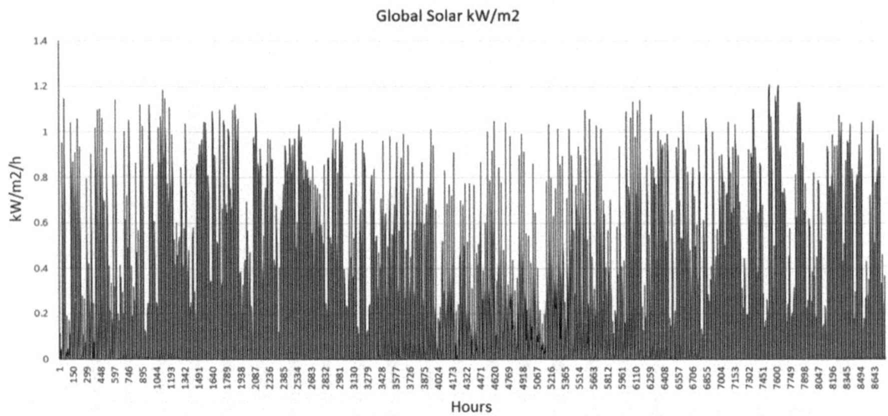

Fig. 2 Global solar radiation in the community under study

2.3 Diesel Generator

The main source of generation is solar panels; however, to provide a high degree of reliability, a diesel generator is used which will be self-sized by the HOMER software. It has a fuel curve intercept of 0.157 L/h, the fuel curve slope is 0.236 L/h/kW, and the cost per kW is 500 USD. Equation 2 is used to determine the amount of fuel consumed, where the capacity of the diesel generator is the power of the generator in a given time, $ab)R_{DG}P_{DG}(t)$:

$$F_{DG}(t) = aR_{DG} + bP_{DG}(t) \tag{2}$$

2.4 Battery Bank

The storage system is the only component of the system that fulfills two functions, that of generating energy and storing in case of excess generation. In Eq. 3, the maximum energy that can be stored is calculated, where kQ_1 is the available energy, is the initial energy stored, is the storage capacity, is the storage constant, and is each storage step. To know the download capacity, use the . The battery used is from the manufacturer kinetic, with a capacity of 140 Ah and the cost of the unit is 210 USD, $Q(t)ck\Delta tQ$max [10]:

$$P_b(t) = \frac{kQ_1(t)e^{-k} + Q(t)kc\left(1 - e^{-k\Delta t}\right)}{1 - e^{-k\Delta t} + c\left(k\Delta t - 1 + e^{-k\Delta t}\right)} \tag{3}$$

The connection between the direct current (DC) bus of the solar panels, the battery bank, and the alternating current (AC) bus is made by an inverter, selected from the manufacturer Fronius, 2.5 kW. The cost per kW is 520 USD.

3 Results and Discussion

3.1 Studio Technical Economic

The optimal sizing is done by the HOMER software comparing the minimum cost of the NPC system and the levelized cost of the COE energy. To calculate the values, Eqs. 4 and 5 are used, where C_{ACC}, C_{ARC}, and C_{AOM} refer to the annualized, replacement, and operation and maintenance costs [11]:

$$\mathrm{NPC} = \frac{C_{\mathrm{ACC}} + C_{\mathrm{ARC}} + C_{\mathrm{AOM}}}{E_{\mathrm{AES}}} \tag{4}$$

The HOMER software calculates the cost of energy by dividing the total generated and usable energy by the annual cost of electricity generation:

$$\mathrm{COE} = \frac{C_{\mathrm{ACC}} + C_{\mathrm{ARC}} + C_{\mathrm{AOM}}}{\mathrm{CRF}\,(i, n)} \tag{5}$$

The levelized cost of energy (COE) is the average cost per kilowatt hour of consumption (kWh) of the energy generated by the system and used by the load. It is the fundamental value in the economic study because it determines the minimum price that energy is produced where all the operating costs of the hybrid energy system are covered. The COE is the crucial value used to compare different power system configurations based on their profitability. Evaluating the value provides fundamental information on economic viability.

The software performs hundreds of simulations to indicate the optimal configuration using the fewest components and lowest energy cost.

3.2 Dispatch Strategies

A dispatch strategy is based on a control algorithm that switches the components according to the requirements of the load. In the literature, there are several control strategies, and among the main ones are cyclic load (CC), load tracking (LF), and combined dispatch. The first two are used for technical-economic studies of the configuration of the system. In the case of LF, the priority has renewable sources avoiding greenhouse gases, and the diesel generator works only if the renewable sources are zero. In each time interval, it is compared with the load, while in the CC strategy [12, 13].

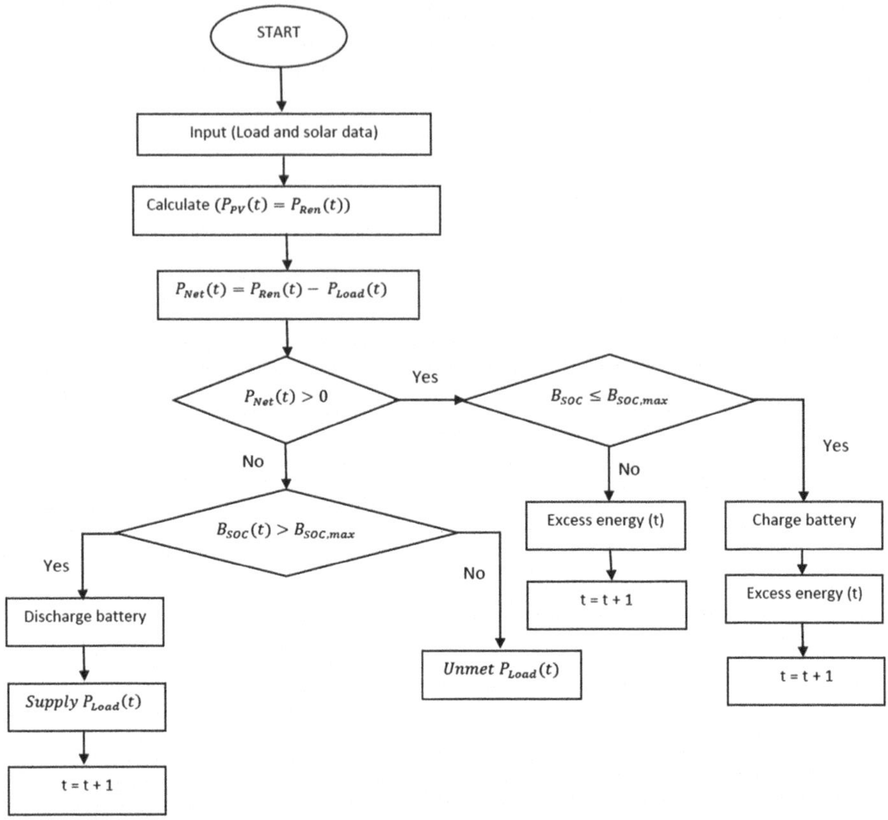

Fig. 3 Load following dispatch strategy

For the study, the load following the strategy has been considered as shown in Fig. 3. In this type of strategy, the main generation source is photovoltaic (PPV(t)) which is compared with the energy demanded by the burden. If there is a surplus of energy (PNet(t) > 0), when the load is satisfied (PLoad (t)), the difference is stored in the battery bank until reaching 100% (BSOC,max). The energy management is important to avoid excess unused energy. In the event that the photovoltaic generator cannot satisfy the load and if the battery energy is available, it will supply the demand; otherwise, the diesel generator supplies the demand that the load demands.

3.3 Results and Discussion

The hybrid system with the LF strategy is made up of solar panels, diesel generator, battery bank, and inverter. In this configuration, the cost of energy is the lowest with 0.322 USD/kWh, as shown in Table 1, while for the DC dispatch strategy, it

Table 1 Energy cost with different dispatch strategies

System optimization results					Dispatch
Architecture	PV (kW)	Generator (kW)	Battery (and)	Inverter (kW)	LF
	10	2.8	10	2.5	
Cost	NPC ($)	COE ($)	Operating cost ($)	Initial capital ($)	
	21,704	0.322	944/year	9500	
Diesel generator	Hours	Production (kWh)	Fuel (L)	O&M ($/year)	
	284	202	92.1	24	
Architecture	PV (kW)	Generator (kW)	Battery (and)	Inverter (kW)	CC
	12	0	16	2.5	
Cost	NPC ($)	COE ($)	Operating cost ($)	Initial capital ($)	
	22,584	0.335	950/year	10,300	
Diesel generator	Hours	Production (kWh)	Fuel (L)	O&M ($/year)	
	0	0	0	0	

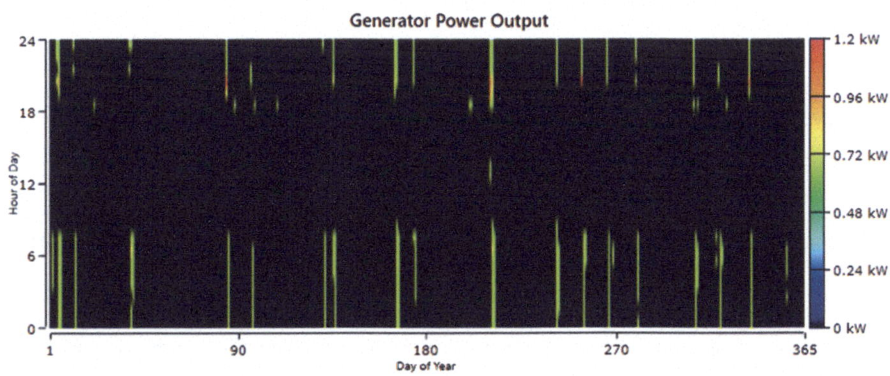

Fig. 4 Diesel generator operation of 2.8 kW

does not use the diesel generator, but the reliability of the system is lower, and the implementation cost may with a value of 22,584 USD.

The resulting configuration with the LF dispatch strategy uses a 10 kW photovoltaic generator that produces 98.6% of the generation which is equivalent to 13,859 kWh/year. The maximum power is 1.58 kW, with an average production of 38 kWh/day. The percentage of penetration is 266% with 4410 h/year. The battery bank is made up of ten units of 140 Ah. The autonomy of the system is 22.3 h; the cost of storage is $0.156/kWh; the nominal capacity is 16.8 kWh; the energy lost by conversion is 452 kWh/year, while that delivered to the load is 2564 kWh/year; and the lifetime is 5.25 years.

The optimization seeks to use the diesel generator for the shortest possible time giving a high degree of reliability. With the optimization, only 1.44% (202 kWh/year) works, with 284 h/year equivalent to 54 starts/year. In Fig. 4, the operation of the generator at night is shown.

4 Conclusions

The electrification of rural communities that are isolated has been carried out in different parts of the world; however, minimizing the cost of energy is a task that is still pending. The optimization of the system reduces the number of components of the hybrid system; however, using the correct dispatch strategy further reduces the cost of energy. In the work, two types of dispatch strategies are compared: the load cycle and the load tracking, the second being the one with the best results. The difference in the implementation of the system is 880 USD. The cost of energy in LF is 0.322 USD/kWh. With the CC dispatch strategy, it has a higher energy cost with 0.335 USD/kWh and the NPC is 22,584 USD.

The optimal configuration is made up of 10 kW of solar panels, a 2.8 kW diesel generator, and a ten-unit battery bank of 140 Ah. The implementation cost is 21,704 USD, the operating cost is 644 USD, and the annual generation is 14,061 kWh. The hybrid system fully satisfies the demanded load.

References

1. Mousavi, S.A., Toopshekan, A., Mehrpooya, M., Delpisheh, M.: Comprehensive exergetic performance assessment and techno-financial optimization of off-grid hybrid renewable configurations with various dispatch strategies and solar tracking systems. Renew. Energy. **210**, 40–63 (2023). https://doi.org/10.1016/J.RENENE.2023.04.018
2. United Nations.: THE 17 GOALS | Sustainable Development, https://sdgs.un.org/goals. Last accessed 2023/12/22
3. Sawle, Y., Gupta, S.C., Bohre, A.K.: Review of hybrid renewable energy systems with comparative analysis of off-grid hybrid system. Renew. Sust. Energ. Rev. **81**, 2217–2235 (2018). https://doi.org/10.1016/J.RSER.2017.06.033
4. Ahmadi, S., Abdi, S.: Application of the hybrid big bang–big crunch algorithm for optimal sizing of a stand-alone hybrid PV/wind/battery system. Sol. Energy. **134**, 366–374 (2016). https://doi.org/10.1016/j.solener.2016.05.019
5. Bahramara, S., Moghaddam, M.P., Haghifam, M.R.: Optimal planning of hybrid renewable energy systems using HOMER: a review. Renew. Sust. Energ. Rev. **62**, 609–620 (2016). https://doi.org/10.1016/J.RSER.2016.05.039
6. Cordero, P.A., Lata García, J., Jurado, F.: Optimization of an Off-Grid Hybrid System Using Lithium Ion Batteries
7. Technical and economic analysis of different configurations of stand-alone hybrid renewable power systems – a case study. Renew. Sust. Energ. Rev. **62**, 941–953 (2016). https://doi.org/10.1016/J.RSER.2016.05.023
8. Vaziri Rad, M.A., Kasaeian, A., Mahian, O.: Evaluation of stand-alone hybrid renewable energy system with excess electricity minimizer predictive dispatch strategy. Energy Convers. Manag. **299**, 117898 (2024). https://doi.org/10.1016/J.ENCONMAN.2023.117898
9. Chaurasia, R., Gairola, S., Pal, Y.: Technical, economic, and environmental performance comparison analysis of a hybrid renewable energy system based on power dispatch strategies. Sustain. Energy Technol. Assess. **53**, 102787 (2022). https://doi.org/10.1016/J.SETA.2022.102787
10. Das, B.K., Tushar, M.S.H.K., Hassan, R.: Techno-economic optimisation of stand-alone hybrid renewable energy systems for concurrently meeting electric and heating demand. Sustain. Cities Soc. **68**, 102763 (2021). https://doi.org/10.1016/J.SCS.2021.102763

11. Lata-García, J., Jurado, F., Fernández-Ramírez, L.M., Sánchez-Sainz, H.: Optimal hydrokinetic turbine location and techno-economic analysis of a hybrid system based on photovoltaic/hydrokinetic/hydrogen/battery. Energy. **159**, 611–620 (2018). https://doi.org/10.1016/J.ENERGY.2018.06.183
12. Medina, J.P., Lata-García, J.: Optimal model of a hybrid electrical system photovoltaic panel/wind turbine/battery bank, considering the feasibility of implementation in isolated areas. J Energy Storage. **36** (2021). https://doi.org/10.1016/j.est.2021.102368
13. Pedro, P.Z., Juan, L.G.: Sensitivity analysis for a hybrid off-grid PV/DG/BATT system for the electrification of rural communities. Diagnostyka. **23**, 1–7 (2022). https://doi.org/10.29354/DIAG/145499

Recent Tools and Their Roles Towards High-Share Renewable Energy in the Climate-Changing World

Joseph Akpan and Oludolapo Olanrewaju

1 Introduction

Achieving a sustainable climate and energy future requires a high share of renewable energy in the current and future energy production and utilisation. On the other hand, careful planning and strategic organisation are required for renewable energy (RE) to be successfully developed and implemented. The existing body of knowledge highlights the significance of all-encompassing renewable energy planning tools for achieving sustainability in the long run [1, 2]. At the same time, other studies have highlighted the significance of a holistic approach that includes geographical breadth, complete resource evaluations, sustainability assessment, energy storage and stakeholders' education [3–6].

In addition, the need to incorporate climate data, earth observation systems and geodetic navigation satellites into energy planning methodologies has been recently emphasised by numerous researchers, notably the work of Akpan and Olanrewaju [6]. This is because inadequate planning could result from data insufficiency, hence resulting in consequences that could lead to inadequate planning as well as unsatisfactory address to climate change issues.

Future climatic projections should inform energy production and consumption plans [7]. Hence, tools for planning and developing renewable energy projects are now expected to embed several features, including climate change indices. Modern energy planning is expected to take climatic factors into account in addition to industrial energy demand and per capita energy use. Consequently, many international energy organisations such as IRENA, IEA and SE4All have now developed tools that incorporate major criteria mentioned above in energy forecasting and planning. The IRENA report in [8] is a good example of a tool

J. Akpan (✉) · O. Olanrewaju
Industrial Engineering, Durban University of Technology, Durban, South Africa
e-mail: 22176142@dut4life.ac.za; oludolapoo@dut.ac.za

L. Chen (ed.), *Trends in Clean Energy Research*, Green Energy and Technology,
https://doi.org/10.1007/978-3-031-67987-2_7

Table 1 Progress in transition to global renewable energy

Energy planning process	2023	2030 targets (1.5°C scenario)	Progress level
Power			
Renewable power capacity	473 GW	1043 GW	Not on track
Solar PV	346 GW/year	578 GW/year	On track
Wind energy	116 GW/year	360 GW/year	Not on track
Hydropower	7 GW/year	28 GW/year	Not on track
Bioenergy	4 GW/year	28 GW/year	Not on track
Geothermal energy	0.2 GW/year	13GW/year	Not on track
Ocean energy power	0.0018 GW/year	13GW/year	Not on track
Total installed RE capacity	3870 GW	11,174 GW	Not on track
RE share in installed capacity	43%	77%	Not on track
RE share in generation	28%	68%	Not on track
Transport and energy storage			
Electric and plug-in hybrid light passenger vehicle stock	40 million	360 million	Not on track
Electrolyser capacity stock	2.9 GW	428 GW	Not on track
			Not on track
Finance and investment			Not on track
Investment in RE power generation	570 USD billion/year	1550 USD billion/year	Not on track
Investment needs for power grids and flexibility	368 USD billion/year	720 USD billion/year	Not on track

Data source—IRENA [8]

that incorporates climate indices like the 1.5 °C scenario in forecasting that adding 1,000 GW of renewable power per year until 2030 is necessary to triple global renewable power capacity to over 11,000 GW. Based on the same IRENA report, the desired share of RE across different distributed energy resources, transport, storage contribution, and the investment requirement needed to reach the 1.5 °C scenario by 2030 is presented in Table 1. In addition, Fig. 1 shows how the trend and projection (both by energy technologies and regional differences) of how the increasing share of RE could follow.

Originating from the World Energy Transitions Outlook (WETO) [11], the IRENA's principal report, the goal of tripling renewable energy capacity as can be seen in Fig. 1, is a significant affirmation of both the work of IRENA and IEA. Hence, reducing costs, having appropriate carbon capture-sequestration options, finding opportunities to develop a hydrogen hub from existing renewable energy potentials, making maintenance choices promptly, investigating alternative energy options, and ensuring long-term sustainability should all be part of the energy production, supply and utilisation profile.

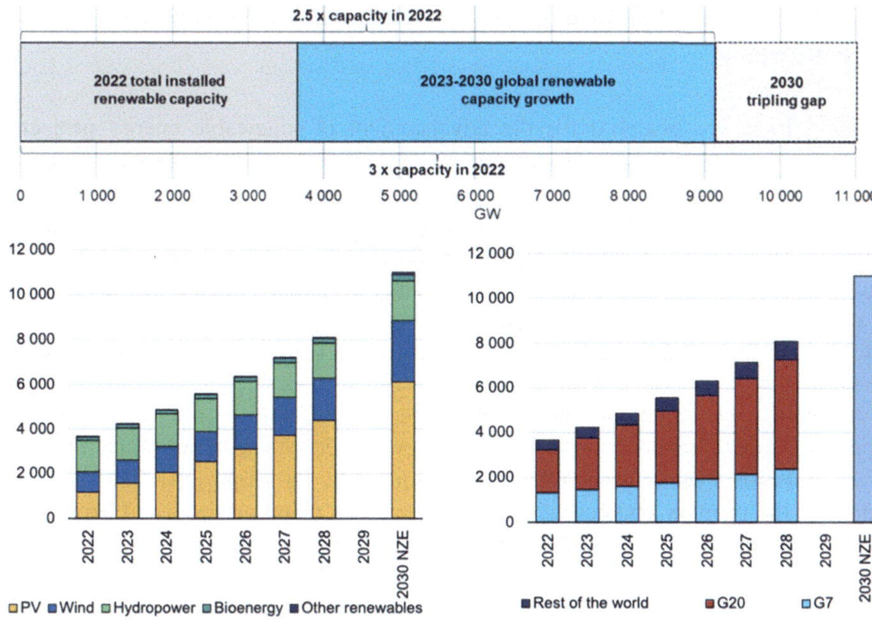

Fig. 1 Trends in RE growth and expectation for net-zero energy society [9, 10]

Therefore, this study looks at comprehensive tools for promoting, planning, developing and coordinating renewable energy projects in line with climate change indices. Here, this study aims to provide an overview of the recent tools adopted in high-share renewable energy planning, considering the established needs for comprehensive resource data, user education and climate-informed strategies for the planning and development of high-share renewable energy projects in a climate-changing world.

The main points of this study are the following:

1. This work discusses the role of planning tool utilisation for the increasing share of RE. It presents a summary of the advantages and limitations of the energy planning process using the tools and then energy planning tools and their capabilities in relation to high-share RE and climate index analysis.
2. The study also discusses recent planning tools with relevant cases used in the planning and development of high-share RE.

This study is structured into five sections. Section 1 presents the introduction of this study; Sect. 2 describes the role of planning tools for the development of high-share RE; Sect. 3, discusses the recent tools; Sect. 4 adds to the tools and relevant case studies; and finally, Sect. 5 concludes this work.

2 The Role of Planning Tools for High-Share Renewable Energy

Planning tools are essential for the advancement of renewable energy projects that hold a significant share of the market. The advantages they offer encompass resource evaluation and site selection optimisation [4], grid integration strategising [12–14], financial projection and investment efficiency [15, 16] and environmental impact evaluation [17]. Hence, these tools enable the identification of the most suitable locations for installing renewable energy systems; simulate the integration of renewable energy sources into the existing power grid; assess project costs, potential earnings and financing options; and reduce the environmental impacts of the project.

A variety of technologies and approaches are available to provide important support towards increasing renewable energy projects. The appropriate tool or set of tools for a particular project is determined by its scope [18]. This section examines the primary characteristics of the selected planning tools and processes for the development of high-share renewable energy to offer valuable insights to the readers.

The tools and approaches in this section are selected based on a recent study that identified several of them as the most suitable for doing 100% RE analysis [6]. The tools vary in their capabilities, ranging from basic energy modelling that provides information for a preliminary study to more advanced tools that enable a comprehensive analysis and provide detailed plans.

Figure 2 presents the key processes involved in the utilisation of the planning tools.

The approach in Fig. 2 is a simplified process for the utilisation of planning tools in developing high RE scenarios. Its advantages and limitations are outlined in Table 2.

Creating a comprehensive list of all goals, together with specifics such as when, where and what is intended to be achieved, is a part of the first step. In the next process, data collection is done which encompasses acquiring details regarding the current energy infrastructure, historical and projected weather patterns and legislation influencing the growth of renewable energy sources.

The defined problem statement is modelled to include economic models for benefit-cost analysis, other simulations and geographic information systems (GIS) for geographical study. Various technological configurations and degrees of renewable energy adoption are considered in the many scenarios laid out by the models.

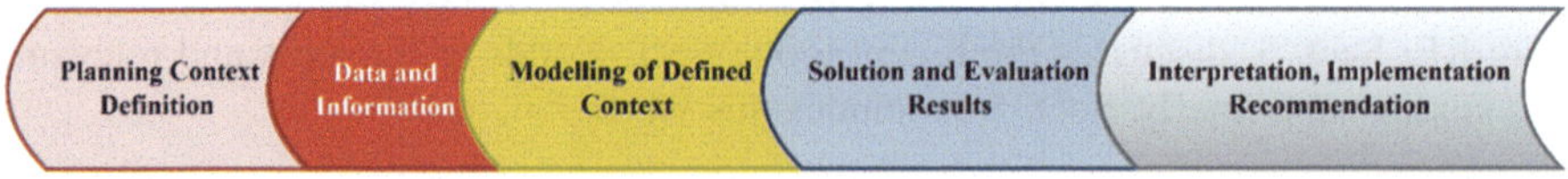

Fig. 2 Simplified process for tool utilisation in high-share RE planning and development

Table 2 Advantages and limitations for tool utilisation in energy planning and development process

Energy planning process	Advantages	Limitation and remarks
Planning context problem definition	1. It makes it easy to have a starting point for planning high-share RE development. 2. There is an opportunity to identify key challenges to overcome in reaching a high share of RE	The problem identification may not be all-encompassing to address the overall sustainability challenges of high-share RE development
Data gathering	1. Unstructured data can be cleaned and restructured to allow for preliminary pattern identification for modelling	This process can be constrained by data accessibility and availability
Modelling of the defined contexts	1. Allows the opportunity to test different hypotheses and scenarios 2. The process can be iterated to obtain the useful results	The modelling process is subject to the user's bias and problem statement as defined within the planning context
Solution and result evaluation	1. The solutions obtained can be evaluated for their suitability in developing policy recommendations and technical solutions	The accuracy of the result is subject to the credibility of input data, and the evaluation process could be subject to user bias
Interpretation, implementation, and recommendation	1. Capable of allowing the techno-economic assessment for decision-making 2. It can provide insight into investment and cost reduction	The full actualisation of interpreted results and policy recommendations are subject to stakeholders' consensus adoption

The outcome of the scenarios is often evaluated in terms of the costs of power generation, reductions in CO_2 emissions, stability of the system with a high percentage of renewable energy and potential social and environmental impacts.

Taking into consideration the objectives and elements made during the definition of the planning context, the model's results are then assessed, and suggested improvements are proposed. In the future where renewable energy sources would be heavily used, planning for high-share renewables becomes very timely and important.

The summary of the studies in Table 3 presents an overview of the primary energy planning tools often used in energy modelling analysis, and their capabilities in assessing a significant proportion of renewable energy and climate indices are highlighted.

The summary of tools presented in Table 3 is conducted in terms of this study's research emphasis on the existing gaps in many studies that often fail to link the high-share RE possibilities with climate change scenarios.

In the next section, the tools that have recently been updated or developed for either resource assessment or incorporation of climate change indices are presented.

Table 3 Summary of major energy planning tools and their capabilities in relation to high-share RE and climate indices analysis

(Country/organisation)	Energy planning tools	Insights for high-share RE	Climate change indices in high-share RE planning scenario	Availability
EU (EU horizon project)	LOCOMOTION with WILLIAMS MEDEAS	Yes	No	https://www.locomotion-h2020.eu/locomotion-models/locomotion-iams/
	EnergyPLAN	Yes	No	https://www.energyplan.eu/
	Mesap PlaNet	Yes	No	https://www.dlr.de/tt/Portaldata/41/Resources/dokumente/institut/system/Scenario_development_with_MESAP.pdf
	HOMER	Yes	No	https://homerenergy.com/
	LOADMATCH	Yes	No	https://www.loadmatch.com/
	TIMES	Yes	No	https://iea-etsap.org/index.php/etsap-tools/model-generators/times
	REMix	Yes	No	https://openenergy-platform.org/factsheets/models/39/
	ISA model	Yes	No	https://isa-specs.readthedocs.io/en/latest/isamodel.html
	PyPSA	Yes	No	https://pypsa.org/
	NEMO	Yes	No	https://www.sei.org/tools/nemo-the-next-energy-modeling-system-for-optimization/
	GENeSYS-MOD	Yes	No	http://www.osemosys.org/genesys-mod.html
Cambridge, USA (climate interactive, the MIT Sloan sustainability initiative and Ventana systems)	VENSIM/C-ROAD/EN-ROAD	Yes	Yes	https://www.climateinteractive.org/c-roads/ https://www.climateinteractive.org/en-roads/
EU (EU horizon project)	RE-INTEGRATE (AU-EU model)	Yes	No	https://re-integrate-au.eu/events/energy-modelling-platform-for-africa-emp-a-2024
Finland (LUT)	LUT ESTM	Yes	Yes	
Middle East (IRENA)	Global Atlas	Yes	No	https://globalatlas.irena.org/workspace
Austria, EU (SE4All)	Universal Integrated Energy Planning (UIEP)	Yes	No	https://www.seforall.org/programmes/universal-integrated-energy-plans
Finland, EU (Wärtsilä)	100% RE Atlas	Yes	Yes	https://www.wartsila.com/energy/towards-100-renewable-energy/atlas-of-100-percent-renewable-energy#/
France, EU (IEA)	IEA energy-climate tool	Yes	Yes	https://www.iea.org/reports/global-energy-and-climate-model
	Energy policy tracker	Yes	–	https://www.energypolicytracker.org/

3 Recent Planning Tools and Their Roles

In this section, information on the capabilities of recent software tools used for high-share renewable energy resources and systems planning and development and with potentials for inclusion of climate change indices are outlined in Fig. 3 and presented in Table 4, alongside comparison with other tools across key element of an ideal planning tool. A follow-up discussion is presented on these tools based on the data found in user manuals and on the websites of programme developers and organisations.

3.1 Global Atlas

An effort by IRENA called the Global Atlas for Renewable Energy (Global Atlas) attempts to help nations that do not have the resources to assess their renewable energy potential and find out what they can do in this area [19]. At the beginning of 2015, 67 nations and more than 50 organisations were involved with the project. The Global Atlas can help narrow users' focus for identifying potential opportunities where additional evaluations could be useful. More than 2000 renewable energy maps are now available on this unified, free online platform, owing to the initiative's efforts, providing a richer dataset for investment planning and development.

Users can overlay data from more than a thousand datasets, allowing them to pinpoint promising locations for more exploration. With the aim of developing the first-ever Global Atlas for Renewable Energy, the tool is collecting maps

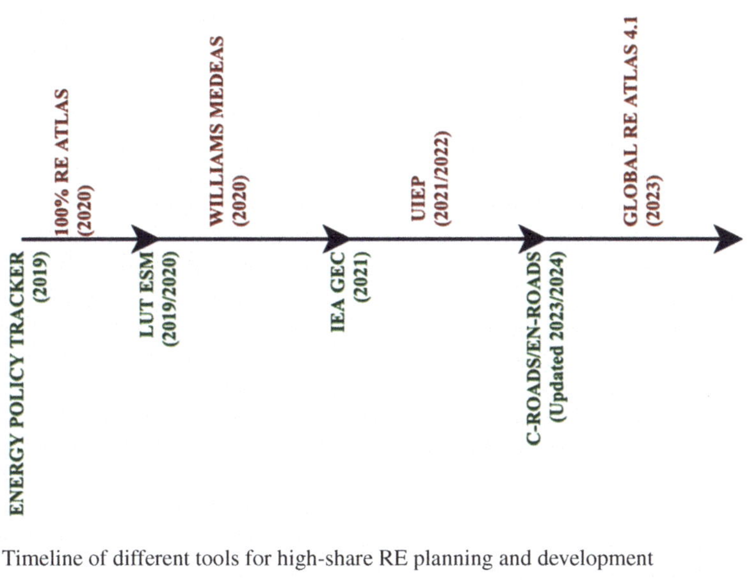

Fig. 3 Timeline of different tools for high-share RE planning and development

Table 4 Modelling capabilities for high-share renewable energy modelling and planning tools

S/N	Key elements	Capabilities	Reviewed tools
1	Data quality	High-quality data for geographical/spatial consideration [1] Representation of both low- and high-emission countries [2] Data fairness [3]	LOCOMOTION with WILLIAMS MEDEAS [2,3] VENSIM/C-ROAD/EN-ROAD [1,3] LUT ESTM [2] Global Atlas [1,2,3] UIEP[1,2,3] 100% RE Atlas [1,2,3] IEA Energy-Climate Tool [1,2,3]
2	Planning	Long-term planning [4] Investment planning tool [5] Generation of policy and regulatory frameworks for case study [6] Transition modelling with representative scenario assumptions [7]	LOCOMOTION with WILLIAMS MEDEAS [4,6,7] VENSIM/C-ROAD/EN-ROAD [4,6,7] LUT ESTM [4,6,7] Global Atlas [4,5] UIEP [4,5] 100% RE Atlas [4,5] IEA Energy-Climate Tool [4,5,6,7]
3	Integration	Carbon capture and storage modelling [8] Off-grid integration [9] Optimisation [10] Inclusion of both energy for electricity and non-electricity purposes [11]	LOCOMOTION with WILLIAMS MEDEAS [9,11] VENSIM/C-ROAD/EN-ROAD [8,11] LUT ESTM [8,10,11] Global Atlas [11] UIEP [9,11] 100% RE Atlas [11] IEA Energy-Climate Tool [8,11]
4	Tool coupling and transparency	Interoperability with existing EST [12] Public transparency of datasets and source codes [13] Compliance with standards [14]	LOCOMOTION with WILLIAMS MEDEAS [12] VENSIM/C-ROAD/EN-ROAD [13,14] LUT ESTM [12, 13,14] Global Atlas [13] UIEP [13,14] 100% RE Atlas [13] IEA Energy-Climate Tool [13,14]
5	100% RE evaluation metrics	Energy system analysis (ESA)[14] Renewable resource assessment (RRA) [15] New technology integration with energy storage requirement (TIESR) [16] Economic, environmental and social impacts (EEI) for sustainability [17] Policy and regulatory analysis (PRA) [18]	LOCOMOTION with WILLIAMS MEDEAS [14,15,18] VENSIM/C-ROAD/EN-ROAD [14,15,16,17,18] LUT ESTM [14,15,16,17] Global Atlas[15] UIEP [15,17] 100% RE Atlas[15,16] IEA Energy-Climate Tool[14,15,16,17,18]

(continued)

Table 4 (continued)

S/N	Key elements	Capabilities	Reviewed tools
6	Forms of energy system analysis	Function Methodologies and mathematical approaches Time horizon Coverage Data requirements Logical assumptions	Present in all the reviewed tools

Key elements and capabilities are adapted from [6]. The numbers 1–18 are nomenclatures used to show the commonality and similarities with each capability from the different key elements of an ideal energy modelling tool

on renewable energy resources from around the world, including solar, wind, geothermal, bioenergy and marine power [19].

3.2 Universal Integrated Energy Planning (UIEP) Tool

Sustainable Energy for All (SEforALL) supports the development of national integrated energy plans (IEPs), which are frameworks that efficiently direct resources to areas where they are needed most. These plans define a country's goals for electrification and clean cooking access and how partners can help them achieve these goals [20].

A well-designed IEP can help a nation achieve its energy goals quickly and affordably. It provides market knowledge to the private sector, policymakers and funders by identifying key geographic areas for extending energy access and implementing cost-effective technologies.

The Universal Integrated Energy Planning (IEP) programme aims to establish the benchmark for superior IEPs and collaborate with partners to ensure countries effectively utilise this tool. Activities include implementing IEPs in countries that need improvement, developing practical tools and knowledge products and offering customised advisory assistance to countries that have established or are in the process of establishing an IEP [20].

The SEforALL initiative has birthed the development of the IEP data standards and IEP for two countries, as follows:

- Malawi Integrated Energy Planning Tool https://malawi-iep.sdg7energyplanning.org/
- Nigeria Integrated Energy Planning Tool https://nigeria-iep.sdg7energyplanning.org/

3.3 100% RE Atlas

The 100% Renewable Energy Atlas by Wärtsilä provides a meticulous and effective plan for establishing energy networks that rely entirely on renewable sources [21]. This blueprint is based on data obtained from Lappeenranta-Lahti University of Technology [22] and Bloomberg New Energy Finance.

The Atlas models energy systems from a theoretical starting point, demonstrating an ideal situation for renewable energy that only depends on hydropower and geothermal sources. Furthermore, it provides the opportunity to incorporate all forms of renewable technologies into the future capacity mix or a scenario that includes solely wind, solar and battery energy storage. The model integrates existing capacity, renewable energy sources and fossil fuels, and the systems are maximised by gradually increasing the utilisation of renewables until reaching 100% RE Systems.

The model includes 145 nations and regions worldwide, considering the sporadic and seasonal characteristics of renewable energy [19]. The data presents the optimal blend of energy sources for a completely sustainable energy system in any particular region, along with the required array of adaptable technologies to facilitate the utilisation of renewable energy. The total capacity of power generation technologies in each country exceeds 100%. However, when considering energy storage, the capacity numbers can surpass 100%.

3.4 Within Limits (WILLIAMS) MEDEAS

The WILIAM Integrated Assessment Model (IAM) is a multiregional model created as part of the LOCOMOTION project [23]. It encompasses nine global areas, which include the EU-27. The initiative commenced in 2005 and normally extends until 2060 in order to achieve net decarbonisation objectives. The model is constructed based on the pre-existing MEDEAS model, which was produced for the MEDEAS project sponsored by the European Union. The study employs many methodologies such as system dynamics, input-output analysis, energy return on investment estimates, life cycle analysis, land and carbon footprinting and microsimulation to analyse intricate relationships between humans and their environment.

3.5 LUT ESM

The LUT Energy System Transition Model (LUT-ESTM) is a tool developed by a Finnish university that combines the electricity, heat, transport and industry sectors into a unified energy system [22]. It encompasses manufacturers of both electricity and thermal energy. The model's optimisation relies on linear optimisation of system parameters while considering restrictions and assuming perfect anticipation of renewable energy output and demand. The objective is to minimise the overall cost of the system, which includes the annualised capital expenditures, operational expenditures, fuel expenses and CO_2 emission costs for all viable technologies [22].

3.6 EN-ROADS/C-ROADS

EN-ROADS/ C-ROADS is a global climate simulator that allows users to assess the impacts of various policies, such as emission reduction, the adoption of electric transportation, implementation of carbon pricing and improvements in agricultural practices, on factors such as energy expenses, temperature, air quality and sea level rise. EN-ROADS is a system dynamics model developed by Climate Interactive, the MIT Sloan Sustainability Initiative and Ventana Systems [24]. The input data is derived from the most consistent scientific data and has been adapted to align with several integrated assessment, climate and energy models. The software functions optimally swiftly accomplishing tasks and is freely accessible online and features a user interface that is intuitive and easy to explore. Furthermore, it has the capability to accommodate over 12 languages.

The tool helps users match their concerns with available solutions to ensure a resilient future by allowing users to quickly evaluate the long-term impacts of global climate policies and activities. The tool provides legislators, educators, businesses, the media and the public with the chance to analyse the likely consequences of energy, economic growth, land use and other policies and uncertainties, with the aim of improving their understanding.

3.7 Energy Policy Tracker

Energy Policy Tracker is a comprehensive tool that provides an exact representation of public funding commitments related to energy production and consumption in G20 countries since the start of the pandemic. The system documented almost 200 distinct policies from these nations, combining the amounts promised in each policy to determine the total aggregate values [25].

The data is divided into five categories: "Fossil unconditional", which advocates for the production and use of fossil fuels without any climate objectives or additional pollution reduction requirements; "Fossil conditional", which supports the production or use of fossil fuels with climate objectives or additional pollution reduction requirements; "Clean unconditional", which promotes low-carbon energy production or use with minimal environmental consequences if proper measures are taken; and "Clean conditional", which supports the transition away from fossil fuels but does not specify the implementation of suitable environmental measures.

Other energy initiatives that fall under this category encompass nuclear energy, first-generation biofuels, biomass and biogas, incineration, hydrogen of unspecified origin and a blend of fossil fuels and clean energy. The categorisation of these policies is defined by their environmental profile, which is established by the energy sources they endorse and whether they are associated with any environmental requirements.

4 Relevant Studies Using the Tools

The evaluated software tools from Section 3 have been utilised by several organisations and researchers in modelling and analysing the future of energy system with policy recommendations. Some of the selected utilisation by five organisations and other researchers is presented in Fig. 4 and Table 5, respectively.

All the models presented in Fig. 4 are based on 2050 yearly projection and on different scenarios. For the IEA model, the net-zero scenario is used [26], 1.5°C climate scenario with net-zero for the IRENA model [11, 19], 100% RE scenario for the STANDFORD Wind Water Sunlight (WWS) [27] and LUT models [28] and 1.5°C climate scenario with primary energy supply from 100% RE for the University of Technology Sydney (UTS) model [29]. The implication in terms of

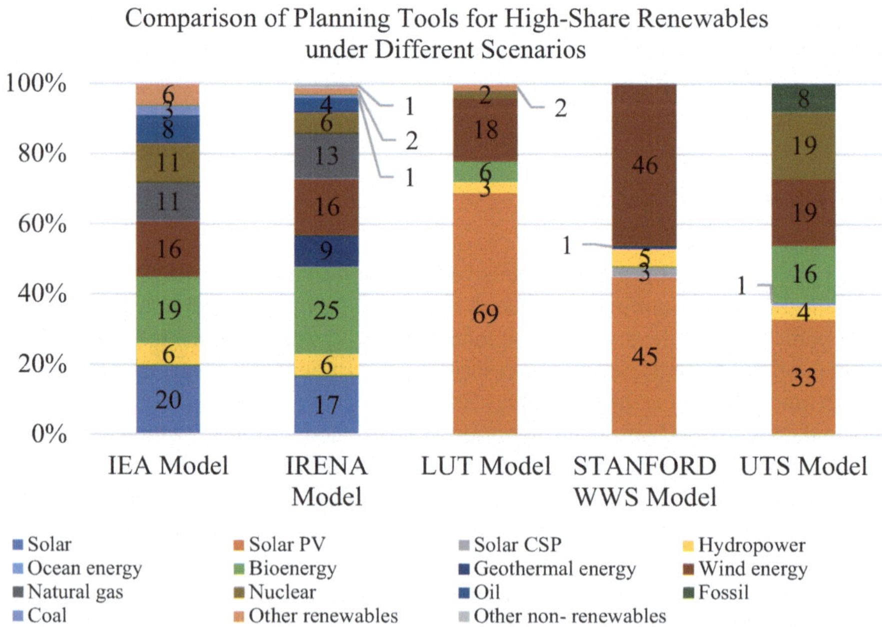

Fig. 4 Comparison of planning tools for high-share renewables under different scenarios

Table 5 Summary of other relevant cases employing the reviewed tool in global high-share RE planning and development

References	Summary of case	Tools employed
[30]	The study looks at nine global transition pathways for decarbonising the energy sector and compares the three best ones in terms of cost, diversity and doing business as usual. Prices of CO_2 and different types of energy are seen as important in all the scenarios. The most optimistic ones show that by 2050, competitive prices will be between 45.2 and 59.2 €/MWh	LUT model
[31]	Using the energy policy tracker, the study determined the disparity between public investment under the 1.5 °C and 2050 scenarios and proposed that an additional 4.47 USD trillion per year—Or around 80% more funding when added to the G20 commitment—Could be used to fund more clean energy projects and initiatives with the goal of keeping global warming within the desired range	Energy policy tracker
[23]	The MEDEAS-world model simulates four variants of current trends, revealing dynamics and implications Absent in classical models. Results suggest continued trends would drive biophysical scarcities, regionalisation, Conflict and global crises, potentially leading to civilisation collapse. This realistic counterfactual scenario Allows for improved sustainable pathways in future work	MEDEAS
[32]	A perspective analysis was carried out to demonstrate how the model links renewable energy share with global average temperature scenarios	VENSIM/C-ROAD/EN-ROAD
[20]	The SEforALL initiative has birthed the development of the IEP data standards and IEP for two countries: • Malawi integrated energy planning tool https://malawi-iep.sdg7energyplanning.org/ • Nigeria integrated energy planning tool https://nigeria-iep.sdg7energyplanning.org/	Universal Integrated Energy Planning (UIEP) planning tool
[21]	With the 100% RE atlas, Wärtsilä estimated the most economical combination of RE sources for Brazil, Australia, Germany, the United Kingdom, the United States and many other nations For Brazil: 36% solar, 25% wind, 36% hydropower, 3% flexible gas and 8% energy storage Australia (east): 29% solar, 64% wind, 4% hydropower, 3% flexible gas and 8% energy storage Germany: 23% solar, 68% wind, 1% hydropower, 8% flexible gas and 13% energy storage United Kingdom: 2% solar, 88% wind, 1% hydropower, 8% flexible gas and 9% energy storage United States (Texas): 37% solar, 57% wind, 7% flexible gas and 13% energy storage	100% RE Atlas

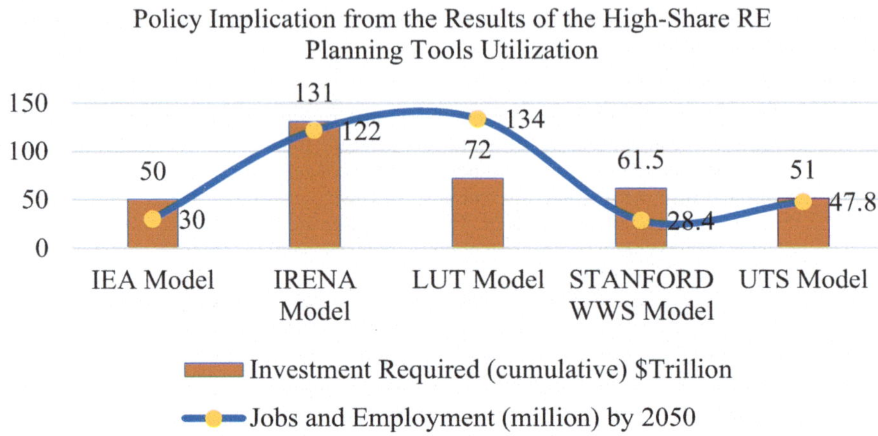

Fig. 5 Policy implication of the tools in high-share RE planning

investment needed and the resulting job creation from the different scenarios of the planning tools shown in Fig. 4 are presented in Fig. 5.

Renewable energy planning is a multifaceted endeavour, with numerous tools available for planning and management for an increasing high share or penetration. In both the IRENA and UTS models where there is similarity in selection of scenario, there still exist variations in the results. For instance, 47.8 million jobs are expected to be created from an investment of $51 trillion with the UTS model. Assuming linearity, and that the amount of $51 trillion is increased by the same factor to $122 trillion, 114.5 million jobs would be created. Hence, a difference of approximately 8 million jobs increase is portrayed from the IRENA model in comparison to the UTS model.

Another variation can be observed in the 100% RE scenarios, when comparing the STANDFORD WWS model with that of LUT. In the STANDFORD WWS model, $61.5 trillion investment yielded 28.4 million jobs. In increasing the initial investment to that of the LUT model, assuming linearity, 61.9 million jobs could be created as against 72 million jobs.

These variations are due to the technology cost differences in the estimated share of the final RE by 2050, as both the LUT and STANDFORD WWS models had different values.

The analysis above indicates how different planning tools and models utilise user-defined configurations to predict energy production and cost and propose transition pathways with policy recommendations that are optimal according to predetermined objectives. In enhancing the high-share RE pathways, it is pertinent to encompass the high-resolution data and the refinement of suitable stakeholders' preferences for obtaining effective results with least variation that will also enhance the management of uncertainty in long-term operation.

In planning and development of high RE share which are often a non-linear process, identifying the most efficient setups could be enhanced using machine learning and cloud computing protocols in fast and more accurate prediction and processing. Such incorporation has the potential to provide valuable insights into the fundamental capabilities of a renewable energy projects while effectively identifying and illustrating the compromises that may need to be made.

Additionally, the insights from the introduction of machine learning and cloud computing technologies could enable users to examine many scenarios and produce comprehensive reports that elucidate the decision-making process and emphasise significant uncertainty enabling a broader stakeholder to comprehend the complexities of renewable energy.

5 Conclusion

This study examines and evaluates recent tools designed for the purpose of planning high levels of RE and provides an analysis of their effectiveness. The research provides analysis of the planning tools' capacity for analysing high-share renewable energy. These tools can enhance performance by utilising several features such as data quality, data fairness, long-term planning, investment planning tools, transition modelling, carbon capture and storage, off-grid integration, optimisation, interoperability, public transparency, compliance with standards, energy system analysis, renewable resource assessment, new technology integration and policy and regulatory analysis. This enhances system cost optimisation and minimises overall RE project expenses. The study encourages the use of the recent tools with enhanced user adaptability for the improvements and to facilitate further research and the deployment of high-share RE.

In order to make well-informed planning and investment decisions, several factors such as highlighted in Tables 2 and 4 are vital for establishing a robust planning for and sustainable energy infrastructure.

Acknowledgement The authors appreciate the support of the Durban University of Technology for funding the publication processing fees.

Conflict of Interest The authors declare no conflict of interest.

References

1. Hansen, K., Breyer, C., Lund, H.: Status and perspectives on 100% renewable energy systems. Energy. **175**, 471–480 (2019, May). https://doi.org/10.1016/j.energy.2019.03.092
2. Asmelash, E., Prakash, G., Gorini, R., Gielen, D.: Role of IRENA for global transition to 100% renewable energy, pp. 51–71 (2020). https://doi.org/10.1007/978-3-030-40738-4_2

3. Marocco, P., Novo, R., Lanzini, A., Mattiazzo, G., Santarelli, M.: Towards 100% renewable energy systems: the role of hydrogen and batteries. J Energy Storage. **57** (2023, January). https://doi.org/10.1016/j.est.2022.106306
4. Derse, O., Yilmaz, E.: Site selection optimization for 100% renewable energy sources. Environ. Sci. Pollut. Res. (2024, March). https://doi.org/10.1007/s11356-024-32733-z
5. Johannsen, R.M., Mathiesen, B.V., Kermeli, K., Crijns-Graus, W., Østergaard, P.A.: Exploring pathways to 100% renewable energy in European industry. Energy. **268** (2023, April). https://doi.org/10.1016/j.energy.2023.126687
6. Akpan, J., Olanrewaju, O.: Towards a common methodology and modelling tool for 100% renewable energy analysis: a review. Energies (Basel). **16**(18), 6598 (2023, September). https://doi.org/10.3390/en16186598
7. Mathiesen, B.V., Lund, H., Karlsson, K.: 100% renewable energy systems, climate mitigation and economic growth. Appl. Energy. **88**(2), 488–501 (2011). https://doi.org/10.1016/j.apenergy.2010.03.001
8. IRENA, Tracking energy transition progress: Tripling renewable power capacity by 2030. The International Renewable Energy Agency. Available: https://www.irena.org/Energy-Transition/Outlook/Tracking-progress
9. IEA, Renewables 2023. 2023. [Online]. Available: https://www.iea.org/reports/renewables-2023
10. IEA, World Energy Outlook 2023, 2023. [Online]. Available: https://www.iea.org/reports/world-energy-outlook-2023
11. IRENA, World Energy Transitions Outlook 2023: 1.5°C Pathway, 2023. https://www.irena.org/Publications/2023/Jun/World-Energy-Transitions-Outlook-2023
12. Hiremath, R.B., Shikha, S., Ravindranath, N.H.: Decentralized energy planning; modeling and application-a review. Renew. Sust. Energ. Rev. **11**(5), 729–752 (2007, June). https://doi.org/10.1016/j.rser.2005.07.005
13. Kriechbaum, L., Scheiber, G., Kienberger, T.: Grid-based multi-energy systems-modelling, assessment, open source modelling frameworks and challenges. Energy Sustain. Soc. **8**, 1 (2018, November 13). https://doi.org/10.1186/s13705-018-0176-x
14. Martinot, E.: Grid integration of renewable energy: flexibility, innovation, and experience. Annu. Rev. Environ. Resour. **41**, 223–251 (2016, October). https://doi.org/10.1146/annurev-environ-110615-085725
15. J. Akpan and O. Olanrewaju, A Novel Evaluation Approach for Emissions Mitigation Budgets & Planning Towards 1.5°C and Alternative Scenarios, Atmosphere (Basel), pp. 1–33, 2024. https://doi.org/10.3390/atmos15020227
16. Yang, Q., Li, Z., Chen, Y., Zhu, Y., Dou, Q.: An investment efficiency evaluation model for distribution network with distributed renewable energy resources. Front. Energy Res. **10** (2022, June). https://doi.org/10.3389/fenrg.2022.931486
17. Akpan, J., Olanrewaju, O., Irusta, R., A state-of-the-art approach for assessing the environmental sustainability of multi-renewable energy systems in the built environment. In: Advances in Clean Energy Systems and Technologies, Green Energy and Technology-Springer Book Series. Springer, Cham (2024). https://doi.org/10.1007/978-3-031-49787-2_31
18. Ciller, P., Lumbreras, S.: Electricity for all the contribution of large-scale planning tools to the energy-access problem. Renew. Sust. Energ. Rev. **120**, 109624 (2020, March). https://doi.org/10.1016/j.rser.2019.109624
19. IRENA: Global Atlas for Renewable Energy, http://irena.org/globalatlas
20. SE4All: Universal Integrated Energy Plans, https://sdg7energyplanning.org/
21. Wärtsilä Energy: Aligning stimulus with energy transformation, 2020. Available: https://www.wartsila.com/insights/whitepaper/aligning-stimulus-with-energy-transformation
22. LUT Energy System Transition Model-brief overview. Available: https://www.greens-efa.eu/files/assets/docs/09._annex_-_accelerating_the_european_renewable_energy_transition_-_2022.pdf
23. Capellán-Pérez, I., et al.: MEDEAS: a new modeling framework integrating global biophysical and socioeconomic constraints. Energy Environ. Sci. **13**(3), 986–1017 (2020). https://doi.org/10.1039/C9EE02627D

24. Climate Interactive: Climate Interactive. Accessed 14 June 2023. [Online]. https://www.climateinteractive.org/
25. IISD, Track public money for energy in recovery packages, 2023. Available: https://www.energypolicytracker.org
26. IEA: Global Energy and Climate Model-Techno-economic Inputs. https://www.iea.org/reports/global-energy-and-climate-model
27. Jacobson, M.Z., Delucchi, M.A., Cameron, M.A., Mathiesen, B.V.: Matching demand with supply at low cost in 139 countries among 20 world regions with 100% intermittent wind, water, and sunlight (WWS) for all purposes. Renew. Energy. **123**, 236–248 (2018, August). https://doi.org/10.1016/j.renene.2018.02.009
28. Bogdanov, D., et al.: Low-cost renewable electricity as the key driver of the global energy transition towards sustainability. Energy. **227**, 120467 (2021, July). https://doi.org/10.1016/j.energy.2021.120467
29. Teske, S.: Achieving the Paris Climate Agreement Goals (Global and Regional 100% Renewable Energy Scenarios with Non-energy GHG Pathways for +1.5°C and +2°C). Springer, Cham (2019). https://doi.org/10.1007/978-3-030-05843-2
30. Aghahosseini, A., et al.: Energy system transition pathways to meet the global electricity demand for ambitious climate targets and cost competitiveness. Appl. Energy. **331** (2023, February). https://doi.org/10.1016/j.apenergy.2022.120401
31. Akpan, J., Olanrewaju, O.: Sustainable energy development: history and recent advances. Energies (Basel). **16**(20), 7049 (2023, October). https://doi.org/10.3390/en16207049
32. J. Akpan and O. Olanrewaju, The Transition to 100% Renewable Energy versus the Global Temperature Scenarios: A Perspective Analysis, in Trends in Clean Energy Research, in Green Energy and Technology-Springer Book Series. Springer, Cham, 2024. https://doi.org/10.1007/978-3-031-67987-2_15

Assessment of Wind and Wave Climate Dynamics in the Mediterranean and Black Seas for Renewable Energy Potential Analysis

Alexandra Ionelia Manolache, Ana-Maria Chirosca, and Liliana Rusu

1 Introduction

The global transition toward renewable energy sources stands as a pivotal mitigation strategy in combating the adverse effects of climate change [1, 2]. Among the diverse renewable energy options, offshore wind and wave energy have emerged as particularly promising due to their abundance and potential for high-energy yields [3]. Recent advancements in technology and infrastructure have further enhanced the feasibility and efficiency of harnessing offshore wind and wave energy, making them increasingly attractive components of the global energy portfolio [4].

Offshore wind energy has witnessed remarkable progress in recent years, with advancements in turbine technology, floating platforms, and installation techniques significantly expanding its reach and potential [5, 6]. Similarly, innovations in wave energy conversion systems have contributed to making wave energy a viable and increasingly competitive renewable energy source [7–9].

At a global level, offshore wind and wave energy resources have garnered considerable attention and investment, with numerous countries leveraging their coastal waters for renewable energy production. The potential of these resources extends beyond individual nations, offering opportunities for international collaboration and resource sharing to maximize energy generation and mitigate climate change impacts [10]. As of 2022, the global capacity for renewable energy stood at approximately 3381.8 GW. Marine energy contributed a mere 0.015% of the total capacity, accounting for about 0.523 GW. Meanwhile, offshore wind energy represented 1.85% of the total capacity, amounting to approximately 62.6 GW [11].

A. I. Manolache · A.-M. Chirosca (✉) · L. Rusu
Faculty of Engineering, Department of Mechanical Engineering,"Dunarea de Jos" University of Galati, Galati, Romania
e-mail: ana.chirosca@ugal.ro

L. Chen (ed.), *Trends in Clean Energy Research*, Green Energy and Technology,
https://doi.org/10.1007/978-3-031-67987-2_8

This highlights the underutilization of marine energy resources compared to other renewable sources, with offshore wind demonstrating substantial growth potential.

With their long coasts and ideal climates, the Mediterranean and Black Seas serve as key hubs for the development of offshore wind and wave energy resources. The Mediterranean area, with its moderate temperatures and strong winds near the coast, is a shining example of the potential for renewable energy. Similarly, the Black Sea region offers significant opportunities for harnessing wind and wave energy, owing to its proximity to densely populated urban centers and growing energy demands. This paper presents the wind and wave climate at key points for renewable energy potential considering the previously mentioned.

2 Materials and Methods

The Black Sea and the Mediterranean Sea have considerable renewable energy potential when it comes to wind energy, but this potential has so far been very little exploited, unlike other European seas.

Aiming to analyze the renewable energy potential, seven points were chosen for the Mediterranean Sea and seven points for the Black Sea, which can be seen in Fig. 1. In addition, the coordinates, water depth, and site description are shown in Table 1.

The methodology outlines the implementation of the wave modeling systems, specifically designed for simulating sea state conditions in the Mediterranean and Black Seas. The SWAN model [12] is used for wave simulations in both the Mediterranean and Black Seas, with consistent parameters validated for both regions [13, 14]. The spatial resolution of the computational domain is 0.2° for

Fig. 1 Locations of selected sites in the Mediterranean and Black Seas

Table 1 Locations of selected sites

Sites	Location	Longitude (°)	Latitude (°)	Depth (m)	Description
M1	Mediterranean Sea	5.12	42.20	−2100.2	Gulf of Lion
M2		−4.57	35.94	−1191.2	Alboran Sea
M3		8.94	38.30	−1934.8	Sardinia
M4		11.82	41.19	−1807.6	Tyrrhenian Sea
M5		17.61	41.09	−1104.8	Adriatic Sea
M6		26.15	35.67	−2325.0	Sea of Crete
M7		32.09	34.52	−2099	Cyprus
B1	Black Sea	37.07	44.14	−2050.9	Russia
B2		31.37	45.62	−44.65	Ukraine
B3		30.30	44.07	−107.32	Romania
B4		28.55	41.76	−86.93	Greece
B5		32.49	42.12	−2152.4	Turkey
B6		37.41	41.76	−1780.2	Turkey
B7		40.56	41.69	−1657.8	Georgia

the Mediterranean and 0.08° for the Black Sea. Given the semi-enclosed nature of these basins, boundary conditions are deemed unnecessary. Input fields for the wave model include only bathymetry and wind data. Therefore, ensuring the accuracy of wind fields is paramount for obtaining realistic results.

Initially, simulations for the historical period (1976–2005) are conducted using high-resolution historical wind fields to force the wave modeling systems. These wind fields, simulated at the Centre National de Recherches Météorologiques (CNRM), France, employ the ALADIN (Aire Limitée Adaptation dynamique Développement InterNational) regional climate model (RCM) [15]. CMIP5 data retrieved from the Copernicus database are utilized, providing comprehensive coverage across Europe with a spatial resolution of 0.11° and a temporal resolution of 6 h [16].

Using the wind fields simulated by the same RCM under various scenarios regarding future greenhouse gas emissions, known as representative concentration pathways (RCPs), simulations of wave conditions in the near future (2041–2070) were carried out with the SWAN model under the scenarios RCP4.5 and RCP8.5.

The next step involves examining wind datasets, assessing wind and wave power, and evaluating climate change impacts.

3 Analysis of Wind and Wave Climate

For the wind climate and wind power density analyses, wind speeds provided by RCM are used, considering the abovementioned scenarios, namely, RCP4.5 and RCP8.5, for the period 2041–2070, together with historical data for the period 1975–2005.

The study aims to assess wind energy potential by analyzing wind power density [17]. This parameter is calculated using wind data obtained at a height of 100 m. The formula utilized for this calculation is provided below:

$$P_w = \frac{1}{2}\rho_{\text{air}} U_{100}^3 \tag{1}$$

where ρ_{air} is the air density and has a value of 1.225 kg/m^3 and U_{100} is the wind speed at a height of 100 m, calculated by a logarithmic law.

The database typically provides wind speed data at a height of 10 m. To assess wind conditions at the height of 100 m, where wind turbines operate, extrapolation is required. This extrapolation can be achieved using a logarithmic law [18], which computes the wind speed at 100 m (U_{100}):

$$U_{100} = U_{10}\frac{\ln(100/z_0)}{\ln(10/z_0)} \tag{2}$$

where U_{10} is the known wind speed at the height of 10 m, while the value of the sea surface roughness length z_0 is taken as 0.0002 m over the sea.

The wave parameters, such as significant wave height (Hs) or wave energy transport (*Etr*), are generated by SWAN model simulations with a 3-h temporal resolution at each node of the computational domain. These time series data are utilized to evaluate wave power over 30-year periods, providing insights into past sea state conditions and future projections under various scenarios. The focus lies on wave power, calculated as the wave energy transport components (Etr), represented as the energy flux per unit of wave crest length (kW/m) [19]:

$$Etr_x = \rho g \iint c_x E(\sigma, \theta)\, dpd\theta \tag{3}$$

$$Etr_y = \rho g \iint c_y E(\sigma, \theta)\, dpd\theta \tag{4}$$

where $E(\sigma,\theta)$ is the directional wave energy density spectrum, x and y are the spherical grid coordinate system (longitude/latitude), and c_x and c_y are the propagation velocities of wave energy in the geographical space (absolute group velocity components).

The total wave power at each grid point is then determined using these components [20]:

$$Etr = \sqrt{Etr_x^2 + Etr_y^2} \tag{5}$$

In Fig. 2, a comparative analysis of the wind speed at 100 m is presented. In each selected point, the mean wind speed is indicated on the maps.

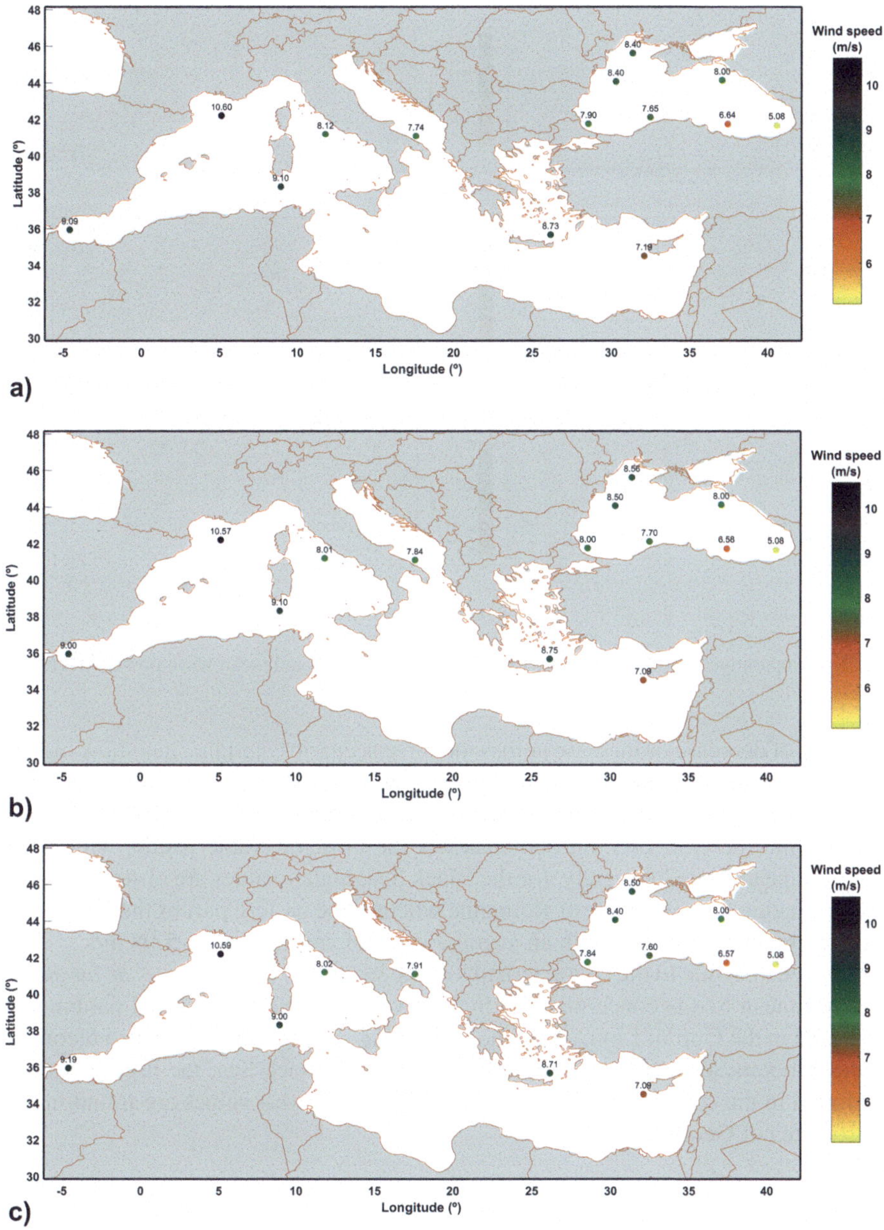

Fig. 2 Mean wind speed. (**a**) Historical, (**b**) RCP4.5, and (**c**) RCP8.5

Upon analyzing the three parameters discussed earlier, we observe that the three studied periods exhibit similar results, with a slight upward trend in values for the near future compared to the recent past. Regarding wind speed, there is no general

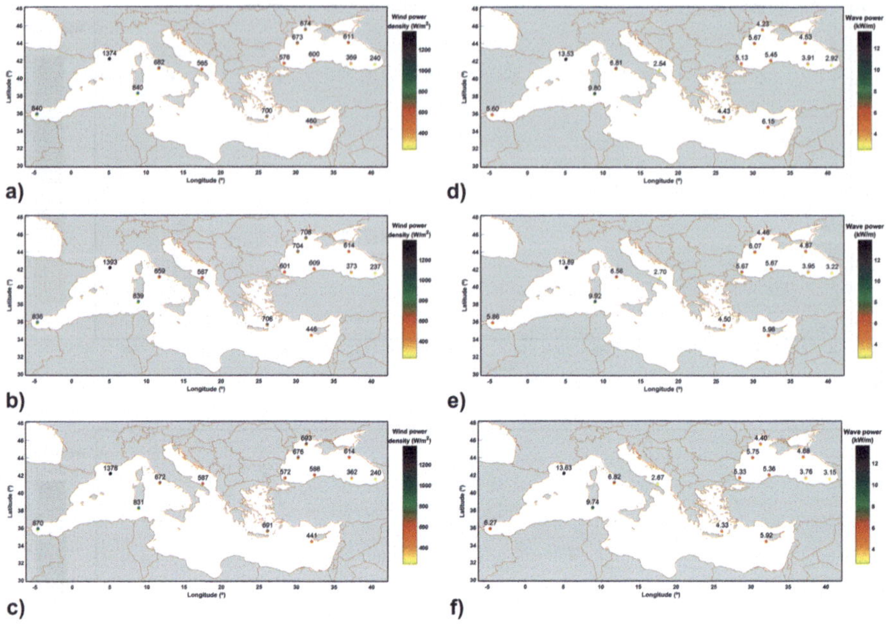

Fig. 3 Comparative analysis of mean wind power density (**a**–**c**) and mean wave power (**d**–**f**)

tendency of decrease or increase across the three scenarios; certain sites show higher wind speed values for the past, while others exhibit higher values for the future.

Regarding the average wind speed value at a height of 100 m (Fig. 2), we observe that the Gulf of Lion demonstrates the most favorable results within the Mediterranean Sea. Conversely, for the Black Sea, higher values are observed in the western region near the coast of Romania, whereas the eastern part of the Black Sea exhibits the lowest results, with an average value of approximately 5.08 m/s.

As expected, due to the characteristics of the wind and wave climate, as far as the Mediterranean Sea is concerned, the highest values of the investigated parameters are found in the Gulf of Lion, as can be seen in Fig. 3. High values are also recorded in the Alboran Sea and near Sardia. As regards the Black Sea, the highest values are found in the coastal areas of Romania, while the lowest values are found in the coastal area of Georgia.

4 Conclusions

Research into wind resource exploitation has become essential to achieving a sustainable energy future, given the global trend toward sustainable energy sources. Hence, the use of renewable energy from offshore wind has been a topic that has attracted particular attention in recent years. Due to the propitious geographical conditions for installing wind farms, as well as the demand for energy to reduce dependence on fossil fuels and combat climate change, the Black Sea and the Mediterranean Sea offer a favorable framework for the development of the wind energy industry.

Currently, although the density of wind farms is not commensurate with the energy potential of these areas, many regions are diversifying their power generation mix in order to make the transition to a sustainable energy system and increase the energetic security of the region.

From the climate analysis presented in this study, the trend for wind speed and wind power is slightly increasing, leading to an increase in wind power potential. Comparison of future projections (2041–2070) with data from the recent past (1976–2005) reveals the impact of global warming and outlines some general trends, among which one of the results of this study can be mentioned, the intensification of winds, and thus the phenomena involving strong winds and storms are expected to be more frequent or to change in both intensity and direction for the Black Sea and the Mediterranean Sea. Thus, to adapt to climate change, it is necessary to strengthen coastal infrastructure, upgrade ships, and better organize ship routes according to problematic areas prone to such events.

Climate change can have a significant impact on sustainable energy development in the Black Sea and Mediterranean regions, but taking action to mitigate its risks will maximize the potential benefits of renewable energy in these regions. The findings suggest that these regions will continue to offer promising conditions for wind energy harnessing and may even offer enhanced opportunities for exploitation.

The development of wind energy exploitation is therefore essential both to combat climate change and to stimulate economic growth, leading to a positive impact on the environment and society.

Acknowledgments This work was carried out in the framework of the research project CLIME-WAR (CLimate change IMpact Evaluation on future WAve conditions at the Regional scale for the Black and Mediterranean seas marine system), supported by the Romanian Executive Agency for Higher Education, Research, Development, and Innovation Funding—UEFISCDI, grant number PN-III-P4-PCE-2021-0015.

References

1. Hassan, Q., Viktor, P., Al-Musawi, T.J., Mahmood Ali, B., Algburi, S., Alzoubi, H.M., Khudhair Al-Jiboory, A., Zuhair Sameen, A., Salman, H.M., Jaszczur, M.: The renewable energy role in the global energy transformations. Renew. Energ. Focus. **48**, 100545 (2024). https://doi.org/10.1016/j.ref.2024.100545
2. Gielen, D., Boshell, F., Saygin, D., Bazilian, M.D., Wagner, N., Gorini, R.: The role of renewable energy in the global energy transformation. Energ. Strat. Rev. **24**, 38–50 (2019). https://doi.org/10.1016/j.esr.2019.01.006
3. Reikard, G., Robertson, B., Bidlot, J.-R.: Combining wave energy with wind and solar: short-term forecasting. Renew. Energy. **81**, 442–456 (2015). https://doi.org/10.1016/j.renene.2015.03.032

4. Diaconita, A.I., Andrei, G., Rusu, E.: Estimation of the tower shape effect on the stress–strain behavior of wind turbines operating under offshore boundary conditions. Inventions. **7** (2022). https://doi.org/10.3390/inventions7010011
5. Onea, F., Diaconita, A., Ganea, D.: Assessment of the black sea high-altitude wind energy. J. Mar. Sci. Eng. **10**, 1463 (2022). https://doi.org/10.3390/jmse10101463
6. Chirosca, A.-M., Rusu, L., Bleoju, A.: Study on wind farms in the North Sea Area. Energy Rep. **8**, 162–168 (2022). https://doi.org/10.1016/j.egyr.2022.10.244
7. Gayathri, R., Chang, J.-Y., Tsai, C.-C., Hsu, T.-W.: Wave energy conversion through oscillating water columns: a review. J. Mar. Sci. Eng. **12**, 342 (2024). https://doi.org/10.3390/jmse12020342
8. Ilyas, A., Kashif, S.A.R., Saqib, M.A., Asad, M.M.: Wave electrical energy systems: implementation, challenges and environmental issues. Renew. Sust. Energ. Rev. **40**, 260–268 (2014). https://doi.org/10.1016/j.rser.2014.07.085
9. Ahamed, R., McKee, K., Howard, I.: Advancements of wave energy converters based on power take off (PTO) systems: a review. Ocean Eng. **204**, 107248 (2020). https://doi.org/10.1016/j.oceaneng.2020.107248
10. Abbass, K., Qasim, M.Z., Song, H., Murshed, M., Mahmood, H., Younis, I.: A Review of the global climate change impacts, adaptation, and sustainable mitigation measures. Environ. Sci. Pollut. Res. **29**, 42539–42559 (2022). https://doi.org/10.1007/s11356-022-19718-6
11. IRENA: 2020 *Renewable Energy Statistics 2023.* International Renewable Energy Agency (IRENA), 2023. ISBN 978–92–9260-537-7
12. Booij, N., Ris, R.C., Holthuijsen, L.H.: A third-generation wave model for coastal regions: 1. Model description and validation. J. Geophys. Res. Oceans. **104**, 7649–7666 (1999). https://doi.org/10.1029/98JC02622
13. Rusu, L.: A projection of the expected wave power in the black sea until the end of the 21st century. Renew. Energy. **160**, 136–147 (2020). https://doi.org/10.1016/j.renene.2020.06.092
14. Rusu, L.: An Analysis of the expected wave conditions in the Mediterranean Sea in the context of global warming. Ocean Eng. (2024)
15. Galabov, V.: On the wave energy potential of the Bulgarian Black Sea Coast, 20 June 2013
16. Cos, J., Doblas-Reyes, F., Jury, M., Marcos, R., Bretonnière, P.-A., Samsó, M.: The Mediterranean climate change hotspot in the CMIP5 and CMIP6 projections. Earth Syst. Dynam. **13**, 321–340 (2022). https://doi.org/10.5194/esd-13-321-2022
17. Rusu, E.: Assessment of the wind power dynamics in the North Sea under climate change conditions. Renew. Energy. **195**, 466–475 (2022). https://doi.org/10.1016/j.renene.2022.06.048
18. Rusu, E., Diaconita, A., Raileanu, A.: An assessment of the wind power dynamics in the European coastal environment. E3S Web Conf. **173**, 01002 (2020). https://doi.org/10.1051/e3sconf/202017301002
19. Liang, B., Fan, F., Yin, Z., Shi, H., Lee, D.: Numerical modelling of the nearshore wave energy resources of Shandong Peninsula, China. Renew. Energy. **57**, 330–338 (2013). https://doi.org/10.1016/j.renene.2013.01.052
20. Rusu, E., Onea, F.: Evaluation of the wind and wave energy along the Caspian Sea. Energy. **50**, 1–14 (2013). https://doi.org/10.1016/j.energy.2012.11.044

Evaluation of the Performance of an Activated Carbon's Family for Biogas Upgrading Using the Adsorption Performance Indicator

Khaled Abou Alfa, Deneb Peredo-Mancilla, Vincent Platel, and Cecile Hort

1 Introduction

The extensive use of fossil fuels in recent decades has reached dangerous levels, increasing greenhouse gas emissions and, therefore, increasing global warming. Thus, they must be transitioned to renewable and sustainable sources to reduce environmental issues [1]. Bioenergy, particularly biogas produced via anaerobic digestion of biomass wastes, is considered a viable alternative energy source [2]. It is a mixture of gases, composed mainly of methane (CH_4) (40–75% v/v) and carbon dioxide (CO_2) (25–55% v/v) with small amounts of other gases such as hydrogen sulfide (H_2S), ammonia (NH_3), water vapor, siloxanes, and halogenated hydrocarbons. Removing or reducing the CO_2 content in biogas is essential to maximize its energy content and make it suitable for various applications, such as natural gas grid, and fuel for vehicles [3]. Various methods have been employed for biogas upgrading to separate the CO_2 content, including pressure swing adsorption (PSA), absorption, membrane separation, cryogenic upgrading, and biological separation [4]. Among the techniques above, PSA is one of the commonly used methods for biogas upgrading; the mechanism behind the PSA involves the affinity and interaction between the CO_2 molecules and the solid's surface under varying pressure conditions. A wide variety of adsorbents have been used to separate CO_2 from biogas mixtures, such as ACs, silica gels, metal-organic frameworks, zeolites, biochars, carbon nanotubes, and carbon molecular sieves [3, 5]. The

K. A. Alfa (✉) · V. Platel · C. Hort
Laboratoire de Thermique, Energetique et Procedes-IPRA, Univ Pau, Pays Adour/E2S UPPA, EA1932, Pau, France
e-mail: kaalfa@univ-pau.fr

D. Peredo-Mancilla
Departamento de Ingeniería en Pesquerías, UniversidadAutónoma de Baja California Sur, La Paz, B.C.S., Mexico

L. Chen (ed.), *Trends in Clean Energy Research*, Green Energy and Technology,
https://doi.org/10.1007/978-3-031-67987-2_9

essential characteristics of the ACs that attracted researchers to choose them for CO_2 adsorption are their excellent chemical and thermal stability, thermal conductivity, less impact to the presence of water, high surface area (up to 4000 $m^2\ g^{-1}$), wide presence of micropores, and availability of low-cost precursors [3].

ACs can be engineered to have greatly varying properties that increase CO_2 adsorption and selectivity for CO_2 over CH_4 by enhancing their physicochemical properties [6]. Many studies have been conducted on increasing the surface basicity of the ACs to improve the adsorption of the CO_2 pure gas and the separation of the CO_2/CH_4 mixture due to the Lewis acid behavior of CO_2 molecules [3, 7]. For instance, introducing basic surface functionalities to the ACs presented a high CO_2 adsorption capacity with more than 4.5 mol kg^{-1} at 0.1 MPa and 298 K under low carbonization and activation conditions [8]. Similar findings were observed by Khalil et al. in which the chemical impregnation of AC with monoethanolamine caused an increase in the adsorption capacity from 18 to 49 mg/mg compared to the untreated AC at atmospheric pressure and room temperature [9]. Regarding the CO_2/CH_4 mixture, Peredo-Mancilla et al. observed that the selectivity ($S(CO_2/CH_4)$) showed the highest value on the AC treated chemically with oxygen and ammonia due to the increased presence of basic surface [7]. The use of ACs to separate a CO_2/CH_4 mixture in the presence of other impurities can significantly influence the separation process in several ways; these gases may compete for adsorption sites on the AC's surface with CO_2 or CH_4. As a result, their presence can affect the adsorption capacity of CO_2 and CH_4 and, therefore, the selectivity. Furthermore, these contaminants can foul the AC's surface, decreasing the available surface area and lowering the AC's overall capacity to capture CO_2 efficiently [10, 11].

In this context, the present work aims to evaluate and compare the adsorption performance of the studied ACs based on a new indicator called "adsorption performance indicator," which comprises the three adsorption parameters (heat of adsorption, working capacity, and selectivity of the CO_2/CH_4 from an equimolar mixture) [12]; the latter two factors were determined and published by Peredo et al. [7].

2 Materials and Experimental Apparatuses

2.1 Materials

Three ACs of the CNR family have been selected for comparative analysis: one commercial (CNR-115) and two modified (CNR-115-ox and CNR-115-ox-am). The commercial AC Norit CNR-115 was purchased from Cabot Corporation company in France which has been modified to improve its efficiency for biogas separation processes. The chemical treatment of commercial CNR-115 was carried out in two successive steps at the Mulhouse Materials Science Institute (IS2M) in France, as

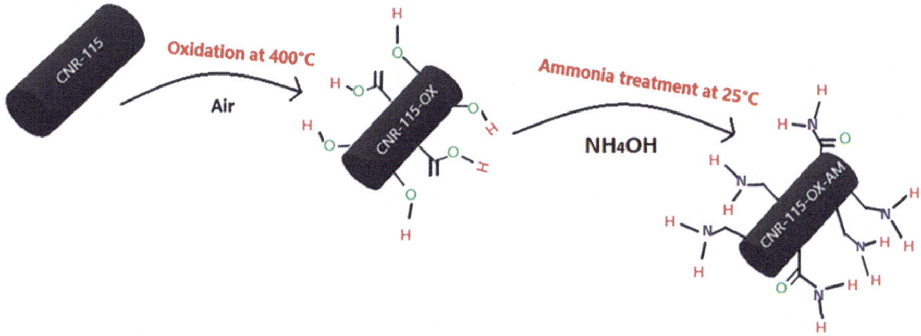

Fig. 1 Chemical treatments of the commercially activated carbon CNR-115

shown in Fig. 1: CNR-115 was oxidized in air at 400 °C for 2 h to increase the amount of oxygenated groups, followed by ammonium vapor (NH_4OH) treatment at 25 °C (298 K) for 36 hours to increase the amount of nitrogen groups [7].

Physicochemical Characterization The textural and chemical characteristics of the three samples were obtained at the Mulhouse Materials Science Institute (IS2M) and the Laboratory of Complexes Fluids and their Reservoirs (LFCR) of the University of Pau and Pays de l'Adour (UPPA) in France and already published by Peredo-Mancilla et al. [7]. All the texture properties of the commercial AC decreased after the two-step chemical modifications. The oxidation step significantly decreased the surface area from 1714 to 929 m^2 g^{-1} which may be attributed to the formation of carbon-oxygen complexes on the surface. Also, the treatment of the oxidized AC with the ammonia vapor decreased the surface area to 352 m^2 g^{-1} because the additional nitrogen groups block some pores. Regarding the chemical characteristics, the oxidation step leads to an increase in the total oxygen-containing groups (such as carboxyls and alcohols) with a value of 1.39 mmol g^{-1} compared to the commercial (0.49 mmol g^{-1}). However, the ammonia treatment promotes the formation of basic surface groups that can be observed in the increase of the pH_{pzc} from 5.87 to 6.48.

2.2 High-Pressure Differential Scanning Calorimeter

The heat of adsorption of CO_2 into the ACs was calculated by integrating the adsorption capacity of CO_2 from the adsorption isotherms and the heat involved, which was measured using a high-pressure differential scanning calorimeter (micro DSC VII from Setaram). The minimum temperature that this calorimeter can reach is about −45 °C, and the maximum temperature is 120 °C, thanks to Peltier thermoelectric devices, with uncertainty estimated at 0.2 K. The calorimetric block is composed of two cells, i.e., the measurement and reference cells (see Fig. 2),

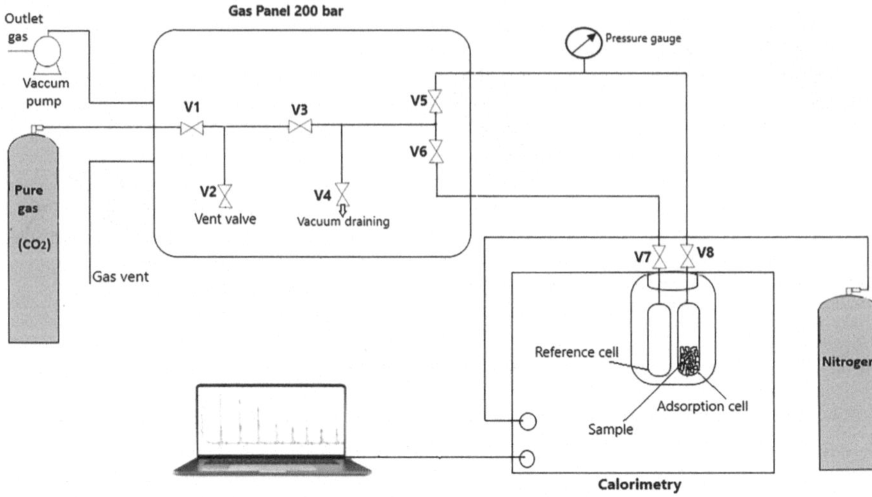

Fig. 2 Schematic of the calorimeter "micro DSC VII"

which are gas-tight high-pressure cells made of stainless steel. The samples were accurately weighed (≈70 mg) before being introduced into the measurement cell. First, to remove any fluid particles contained in the system, a regeneration process takes place by subjecting the sample to a couple of vacuum (10^{-2} Pa) (through the gas panel) and heating at a temperature of 100 °C for 24 h. Afterward, the temperature in the furnace was kept at 30 °C during the overall adsorption process to ensure uniform temperature distribution and stabilization of the heat flow signals. A quantity of the pure gas (CO_2) was introduced and regulated through the gas panel to the calorimetry until the required pressure is reached.

2.3 Determination of Adsorption Performance Indicator (API)

The separation of a constituent from a gas mixture using adsorbents is based on three primary parameters: the selectivity, the adsorption capacity, and the heat of adsorption. Thus, these latter parameters solely cannot describe the performance of the different materials in separating a mixture of gases. Therefore, for the primary selection of the adsorbents for a specific separation process, a single indicator called "adsorption performance indicator" was proposed by Wiersum et al. to combine the three parameters mentioned above with different relative importance of each one depending on the objectives of the study [12]. An adsorbent that has a higher selectivity and adsorption capacity together with a lower heat of adsorption is considered the ideal material for biogas separation, but it is rare to find one that combines these three properties at the same time. The following form can present the API relation:

$$\mathrm{API} = \frac{\left(S_{1/2} - 1\right)^{\mathrm{A}} \mathrm{WC}_1^{\mathrm{B}}}{\left|\Delta H_{\mathrm{ads},1}\right|^{\mathrm{C}}} \tag{1}$$

In Eq. (1), $S_{1/2}$ is the selectivity of component 1 over component 2, WC_1 is the working capacity of the stronger adsorbate component 1, and $\Delta H_{\mathrm{ads},1}$ is the heat of adsorption of component 1 which takes place in the denominator of the relation because it is detrimental to the performance of the process. A, B, and C are the exponents of the selectivity, the working capacity, and the heat of adsorption, respectively.

3 Results and Discussions

3.1 Heat of Adsorption

The heat of adsorption indicates the interaction strength between the adsorbed gas and the adsorbent surface, which explains the ACs' affinity to the gas molecules. Figure 3 shows the heat of adsorption calculated by combining the results of the adsorption isotherms of the CO_2 pure gas and the calorimetry measurements of the CO_2 adsorption into the CNR family at 30 °C as a function of pressure. It was shown that the calculated heat of adsorption for all the ACs studied varied in the range of physisorption (<40 kJ mol^{-1}); hence, a reversible adsorption process is expected. The heat of adsorption of CNR-115, CNR-115-ox, and CNR-115-ox-am decreases from 14.95 ± 0.67, 26.78 ± 0.45, and 28.08 ± 0.26 kJ mol^{-1} to 10.01 ± 0.10, 23.29 ± 0.27, and 24.57 ± 1.64 kJ mol^{-1}, respectively, in a corresponding pressure range of 0.1–2.4 MPa. For all the ACs, the heat of adsorption is high in the low-pressure region due to the strong interaction between CO_2 molecules and the favorable active sites of the ACs (monolayer adsorption). As pressure increases, the heat of adsorption typically begins to decrease from its initial high value because more CO_2 molecules are adsorbed onto the AC's surface, and the surface sites become occupied by CO_2, weakening the interactions between the adsorbent and adsorbate. At a certain point, when most of the available adsorption sites are filled (monolayer coverage), the heat of adsorption reaches a plateau in which an additional pressure increase does not significantly alter the heat of adsorption. For the ACs, CNR-115-ox and CNR-115-ox-am, at very high pressures, the heat of adsorption increased due to the electrostatic interactions between CO_2 and the chemical functional groups on the material's surface. The heat of adsorption of CO_2 on commercial AC (unmodified) is often lower than on chemically modified AC because it primarily relies on physical adsorption

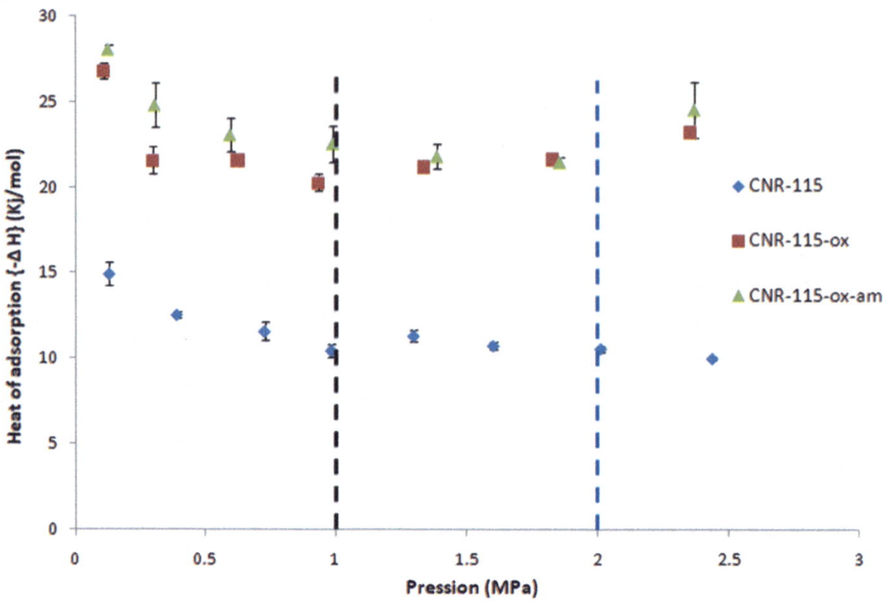

Fig. 3 Heat of adsorption of the activated carbons of the CNR-115 family at 30 °C as a function of pressure

mechanisms, such as van der Waals forces, and these interactions are generally weak. Regarding the regeneration energy requirements, CNR-115-ox-am and CNR-115-ox require higher energy to break the strong adsorbate-adsorbent interactions, but the dominance of the physisorption range can mean that the CNR-115-ox-am is still regenerable [7].

3.2 Adsorption Isotherms

This part includes the results obtained of the equimolar CH_4/CO_2 separation into the ACs's family by Peredo et al. for the adsorption capacity and the selectivity at 30 °C and 1 and 2 MPa [7]. The adsorption of CO_2 in the oxidized CNR-115-ox is higher than in the unmodified sample due to the presence of oxygen functionalities of groups, which induce the formation of hydrogen bonds between CO_2 and the carbon surface. However, this adsorption capacity decreased after the ammonia treatment due to the pore blockage by the presence of the new nitrogen surface groups. Regarding the selectivity, which is the preferential adsorption of CO_2 over CH_4, it was seen that the three ACs had a high selectivity as it is more than 1. At 2 MPa, CNR-115-ox-am displayed the highest selectivity ($S(CO_2/CH_4)$) of 102 ± 1 compared to the two other ACs. This is due to its basic surface modification, which enhances the CO_2 uptake with a negligible CH_4 adsorption.

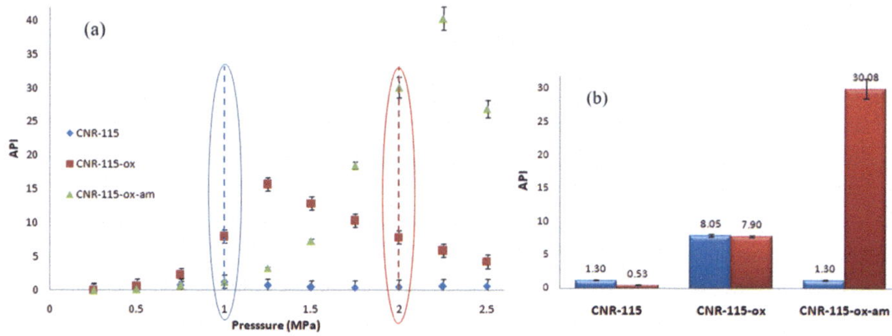

Fig. 4 (**a**) API variation as a function of pressure. (**b**) API values of the CNR family at 1 MPa (blue) and 2 MPa (red)

3.3 Adsorption Performance Indicator

In this study, API is used to evaluate the performance of the three adsorbents for removing CO_2 molecules from CH_4 molecules. In this regard, we give higher importance to the parameter "adsorption capacity" with an exponent "B = 2," but the exponents (A and C; see Eq. 1) of the two remaining parameters are kept to 1. Figure 4a presents the API variation as a pressure function, API results indicate that the ideal adsorbent differs by varying the pressure; under the low-pressure region (<1.5 MPa), the adsorbent CNR-115-ox has the highest values with a maximum of 15.74 ± 0.48 at 1.25 MPa, which means that the presence of oxygen groups enhances the separation of CO_2/CH_4 in this region. However, by increasing the pressure above 1.5 MPa, API suggests that CNR-115-ox-am is the preferred adsorbent with a maximum value of 40.27 ± 1.69 at 2.25 MPa and is likely to perform much better than the two other adsorbents. This is due to its higher selectivity under this pressure region, thanks to the electrostatic interactions between the CO_2 molecules and the basic nitrogen surface. Figure 4b shows the API values of each adsorbent at the two pressures, 1 and 2 MPa; at 1 MPa, the adsorbent CNR-115-ox has the highest value (8.05 ± 0.24) compared to the other two adsorbents as a result of the presence of oxygen groups. At 2 MPa, CNR-115-ox-am is identified as the most effective adsorbent, achieving a selectivity of 30.08 ± 1.5, attributed to its basic nitrogen surface. The sequence order of the heat of adsorption, the working capacity of CO_2, and the selectivity of CO_2 over CH_4 on these adsorbents does not correspond to the order of the API values. This demonstrates the significance of the API indicator in screening ad sorbents for binary gas separations.

4 Conclusions

Three ACs of the CNR family, one commercial (CNR-115) and two modified (CNR-115-ox and CNR-115-ox-am), were used to evaluate their performance on separating the equimolar CO_2/CH_4 mixture at 30 °C. The evaluation of this performance was done using the adsorption performance indicator, which integrated the three main adsorption parameters: the heat of adsorption in which the three samples are within the range of physisorption, but the modified carbons have higher values due to the chemical interactions with the CO_2 molecules. The CO_2 working capacity in which the oxidized CNR-115-ox has the highest values due to the presence of the oxygen groups, the selectivity ($S(CO_2/CH_4)$) in which the more selective is the CNR-115-ox-am at higher pressure due to the chemical interaction of the CO_2 molecules with the basic surface, with a negligible CH_4 adsorption. Finally, API results indicate that the adsorbent CNR-115-ox is preferred over the two other adsorbents under the low-pressure region as a result of the presence of the oxygen surface groups, but CNR-115-ox-am is the most promising carbon material above 1.5 MPa due to its higher selectivity as a result of the presence of the nitrogen functional groups.

Acknowledgements This research was carried within the E2S UPPA framework, supported by the "Investissements d'Avenir" French program managed by ANR (ANR-16-IDEX-0002).

References

1. Abbasi, T., Abbasi, S.: Is the use of renewable energy sources an answer to the problems of global warming and pollution? Crit. Rev. Environ. Sci. Technol. **42**(2), 99–154 (2012)
2. Guo, M., Song, W., Buhain, J.: Bioenergy and biofuels: history, status, and perspective. Renew. Sust. Energ. Rev. **42**, 712–725 (2015)
3. Abd, A.A., Othman, M.R., Kim, J.: A review on application of activated carbons for carbon dioxide capture: present performance, preparation, and surface modification for further improvement. Environ. Sci. Pollut. Res. **28**(32), 43329–43364 (2021)
4. Petersson, A., Wellinger, A.: Biogas upgrading technologies—developments and innovations. IEA Bioenergy. **20**, 1–19 (2009)
5. Abd, A.A., Naji, S.Z., Hashim, A.S., Othman, M.R.: Carbon dioxide removal through physical adsorption using carbonaceous and non-carbonaceous adsorbents: a review. J. Environ. Chem. Eng. **8**(5), 104142 (2020)
6. Bedia, J., Peñas-Garzón, M., Gómez-Avilés, A., Rodriguez, J.J., Belver, C.: A review on the synthesis and characterization of biomass-derived carbons for adsorption of emerging contaminants from water. J. Carbon Res. **4**(4), 63 (2018)
7. Peredo-Mancilla, D., Ghimbeu, C., Réty, B., Ho, B.N., Perez, D., Vaulot, C., Hort, C., Bessieres, D.: Surface-modified activated carbon with a superior CH4/CO2 adsorption selectivity for the biogas upgrading process. Ind. Eng. Chem. Res. **61** (2022)
8. To, J.W.F., He, J., Mei, J., Haghpanah, R., Chen, Z., Kurosawa, T., Chen, S., Bae, W.G., Pan, L., Tok, J.B.H., Wilcox, J., Bao, Z.: Hierarchical N-doped carbon as CO2 adsorbent with high CO2 selectivity from rationally designed polypyrrole precursor. J. Am. Chem. Soc. **138**(3), 1001–1009 (2016)

9. Khalil, S.H., Aroua, M.K., Daud, W.M.A.W.: Study on the improvement of the capacity of amine-impregnated commercial activated carbon beds for CO2 adsorbing. Chem. Eng. J. **183**, 15–20 (2012)
10. Durán, I., Álvarez-Gutiérrez, N., Rubiera, F., Pevida, C.: Biogas purification by means of adsorption on pine sawdust-based activated carbon: impact of water vapor. Chem. Eng. J. **353**, 197–207 (2018)
11. De Oliveira, L.H., Meneguin, J.G., Pereira, M.V., Nascimento, J.F., Arroyo, P.A.: Adsorption of hydrogen sulfide, carbon dioxide, methane, and their mixtures on activated carbon. Chem. Eng. Commun. **206**, 1–21 (2019)
12. Wiersum, A.D., Chang, J.S., Serre, C., Llewellyn, P.L.: An adsorbent performance indicator as a first step evaluation of novel sorbents for gas separations: application to metal-organic frameworks. Langmuir. **29**(10), 3301–3309 (2013)

Estimation of Power Consumption in Screw Gas Compressor Installed on Natural Gas Power Plant

Piyanart Sommani, Suntiti Yoomak, and Santipont Ananwattanaporn

1 Introduction

Power generation from natural gas has gained significant amount of the proportion in the share of electricity generation source [1]. The reason is natural gas is cleaner than other fossil fuels because it emits the least amount of carbon dioxide into the air when combusted [2]. In addition, natural gas is a flexible fuel, which can be transported in different ways from the source to the power generation units, thereby providing a country with limited resource to secure the energy source [3].

The trend of rising demand for natural gas to secure energy security results in requirement for infrastructure such as natural gas power plants [4]. The initial engineering design phase of the power plant core component such as industrial device, equipment needs estimation of the power requirement, which will be used as a performance index in an aspect of energy efficiency and effect the cost of construction. On the other hand, there is a limitation in the initial design process which requires specification data from the manufacturer such as power consumption under different operating conditions. Thus, the estimation model can simplify this process and can provide benefit in terms of efficient the initial design phase of the plant.

One of the key components of natural gas power plants is gas compressors, which increase the pressure of a gas by reducing its volume [5]. However, there are many types of compressors depending primarily on the required flow to be compressed, the density of the gas in conjunction with the total head, and the duty that has to be performed [6]. The screw gas compressor has gained a lot of attention due to its high efficiency and reliability compared with other types of compressors. In

P. Sommani · S. Yoomak (✉) · S. Ananwattanaporn
School of Engineering, King Mongkut's Institute of Technology Ladkrabang, Bangkok, Thailand
e-mail: suntiti.yo@kmitl.ac.th

L. Chen (ed.), *Trends in Clean Energy Research*, Green Energy and Technology,
https://doi.org/10.1007/978-3-031-67987-2_10

addition, it also has low carbon footprint that will help in reducing the impact on the environment [7].

The literature review of the past research with regard to the performance evaluation of the screw compressors has been done. The study on the characteristics of the single screw compressor presented the diagram ratio between pressure and volume obtained from experimental unit under different conditions, which can provide insights into the relationship between parameters in the compressor [8]. In aspect of evaluating the performance of the single-screw compressor, evaluation indices such as maximum volumetric efficiency, shaft efficiency, irreversible loss rate, and cooling efficiency under operating conditions have been used to analyze compressor performance [9]. For the efficiency of the screw compressor, the methodology based on numerical analysis has been presented. The result can be presented using many indices such as gas and oil mass flow variations, power consumption, and torque value [10]. The energy efficiency and energy consumption of the screw compressor can be determined by using the theoretical discharge pressure that is obtained in the compressor independently of the system discharge pressure [11]. In terms of the natural gas power plants, the data envelopment analysis approach has been proposed with multiple inputs and outputs and using a wide range of data such as the age of a plant and the total number of hours that a plant is in operation during each year [12].

This research aims to create an estimation model for the power consumption in the screw gas compressor in natural gas power plants based on the provided specification and internal parameter equation. The estimated result can be used to provide an accurate initial performance of the screw gas compressor before the initial construction of the power plant in order to select the most energy-efficient screw gas compressor for the project, which then provides the project owner with a long-term low-energy cost and satisfies energy performance.

2 Power Estimation Parameter

In order to accurately estimate the power consumption of the screw gas compressor, the internal parameter of the screw gas compressor that is related to the power consumption of the unit must be identified and calculated.

2.1 Isentropic Compression Work

The work for isentropic compression for screw gas compressors (W_{isen}) can be calculated by using Eq. (1):

$$W_{\text{isen}} = \frac{kR\,(T_S + 273)}{k-1}\left[\left(\frac{P_D}{P_S}\right)^{\frac{k-1}{k}} - 1\right] \tag{1}$$

where:

W_{isen} = ideal work for isentropic compression (J/mol)
k = ratio of specific heat,
R = gas constant 8.314 J/mol.K.
T_S = compressor input temperature (°C).
P_S = compressor input pressure (barA).
P_D = compressor output pressure (barA).

2.2 *Isentropic Compression Efficiency*

The efficiency of isentropic compression for screw gas compressors (η_{isen}) can be calculated by using the correlation between the ratio of mass flow rate and isentropic efficiency as shown in Table. 1.

From the table, the polynomial trendline equation has been created, and the constant value on each order of the polynomial equation can be created as shown in Eq. (2):

$$\eta_{\text{isen}} = 7.5291e^{-5}X^3 - 2.0575e^{-2}X^2 + 1.8943X + 21.076 \tag{2}$$

where:

η_{isen} = efficiency of isentropic compression (%)
X = ratio of mass flow rate (%).

Table 1 Isentropic compression efficiency at different mass flow rates

Mass flow rate (%)	Isentropic efficiency (%)
40	68.75
45	71.42
50	73.88
55	75.63
60	76.96
65	77.89
70	78.60
75	79.12
80	79.53
85	79.73
90	79.84
95	79.94
100	80.00

2.3 Actual Work for Isentropic Compression

The actual work for isentropic compression for screw gas compressors (W_{act}) can be calculated by using Eq. (3):

$$W_{act} = \frac{W_{isen}}{\eta_{isen}} \tag{3}$$

where:

W_{act} = actual work for isentropic compression (J/mol).

2.4 Mechanical Efficiency of Screw Gas Compressor

The mechanical efficiency of the screw gas compressor (η_{comp}) can be calculated by using the polynomial trendline equation, which derived the coefficient from the research by Brasz [12], which studies the continuous operation of the screw gas compressor under different operating conditions. Thus the coefficient for each order under different percentage heads can be summarized as shown in Table 2 and Eq. (4).

$$\eta_{comp} = A3X^3 + A2e^{-2}X^2 + A1X + A0 \tag{4}$$

Table 2 Coefficient value of the polynomial trendline equation for mechanical efficiency

	Coefficient			
Percentage head (%)	A3 (x^3)	A2 (x^2)	A1 (x^1)	A0 (x^0)
0	0.00	0.00	0.00	0.00
50	$3.70e^{-5}$	$-1.44e^{-2}$	1.81	5.68
60	$4.53e^{-5}$	$-1.49e^{-2}$	1.81	9.54
70	$4.73e^{-5}$	$-1.33e^{-2}$	1.66	13.32
80	$3.10e^{-5}$	$-1.06e^{-2}$	1.55	15.37
90	$4.34e^{-5}$	$-1.12e^{-2}$	1.52	16.66
100	$2.38e^{-5}$	$-0.77e^{-2}$	1.35	18.47
110	$1.23e^{-5}$	$-0.55e^{-2}$	1.24	19.37
120	$0.42e^{-5}$	$-0.39e^{-2}$	1.16	19.43

2.5 Actual Work on Screw Gas Compressor Shaft

An actual work on the screw gas compressor shaft (W_{shaft}) can be calculated by using Eq. (5):

$$W_{\text{shaft}} = \frac{W_{\text{comp}}}{\eta_{\text{comp}}} \tag{5}$$

where:

W_{shaft} = actual work at the compressor shaft (J/mol)
W_{comp} = ideal work at the compressor shaft (J/mol).

2.6 Brake Horsepower

Brake horsepower for the screw gas compressor (BHP) can be calculated by using Eq. (6):

$$\text{BHP} = \frac{m \times W_{\text{shaft}}}{3600 \times \text{MW}} \tag{6}$$

where:

BHP = mechanical efficiency of the compressor (%)
m = ideal work at the compressor shaft (J/mol)
MW = molecular weight (g/mol).

2.7 Motor Efficiency

The efficiency of the motor for the screw gas compressor (η_{motor}) can be calculated by using Eq. (7):

$$\begin{aligned}\eta_{\text{motor}} = & -1.5238e^{-10}X^6 + 8.4571e^{-8}X^5 - 1.9341e^{-5}X^4 + 2.3321e^{-3}X^3 \\ & - 1.5646e^{-1}X^2 + 5.5383X + 15.57\end{aligned} \tag{7}$$

where:

η_{motor} = efficiency of motor for the screw gas compressor (%).

2.8 Screw Gas Compressor Power Consumption

Power consumption of the screw gas compressor ($P_{comp,\,tem}$) can be calculated by using ratio between motor efficiency and Eq. (8).

$$P_{comp,tem}\,(kW) = \frac{BHP}{\eta_{motor}} \tag{8}$$

3 Result

To estimate the power consumption, the internal parameter must be obtained from either specification from the manufacturer or measurement data. The main parameters consist of mass flow rate, max mass flow rate, suction pressure, discharge pressure, suction temperature, molecular weight, specific heat ratio, minimum suction pressure, and maximum discharge pressure. After inputting the required parameter in the model, the model can be estimated by the input parameter into the equation and calculate the required value according to Eqs. 1, 2, 3, 4, 5, 6, 7, and 8.

The verification of the proposed model has been done by using data from the actual screw gas compressor that has been installed on the actual natural gas power plants from two different projects (A and B). The operating condition of each screw gas compressor is also taken into consideration with 11 actual scenarios that are obtained from plant operators. The estimated result for each project can be summarized as shown in Tables 3 and 4.

The accuracy of the model can be determined by calculating an error of the screw gas compressor power consumption based on the estimation model in comparison with the data from the actual units that have been installed in the natural power plant in Thailand. An estimated error under different operating conditions is displayed in Table 5.

From the table, the comparative calculated result from the estimated model and actual power consumption under 11 operating conditions has demonstrated that overall, the error is less than 5%. However, this percentage error is still acceptable for the utility because the value is approximated to provide specification.

4 Conclusion

This paper aims to present an estimation model for the power consumption in screw gas compressors. The methodology is based on an internal parameter of the component. The verification process of the proposed estimation model has been done by a comparative study with two actual units installed on the natural gas

Table 3 Estimation value using the proposed model of the screw compressor in project A

Project A		Operating condition				
Parameter	Unit	Off-peak 16.05	Off-peak 17.05	Off-peak 15.40	Off-peak 16.40	Off-peak 17.40
Mass flow rate	kg/h	15,462.0	15,562.8	19,980.0	20,127.6	19,893.6
Max mass flow rate	kg/h	27,000	27,000	27,000	27,000	27,000
Percentage mass flow rate	%	57.27	57.64	74.00	74.55	73.68
Suction pressure	barA	20.99	21.14	21.09	20.99	21.14
Discharge pressure	barA	29.10	28.75	28.93	28.74	29.27
Pressure ratio	–	1.39	1.36	1.37	1.37	1.38
Percentage head	%	92.15	86.48	89.09	88.07	92.39
Suction temperature	°C	20.60	20.44	19.53	19.31	19.62
Molecular weight	g/mol	18.278	18.278	18.278	18.278	18.278
Specific heat ratio	–	1.4007	1.4019	1.4024	1.402	1.4027
Min suction pressure	barA	21.2	21.2	21.2	21.2	21.2
Max discharge pressure	barA	30.0	30.0	30.0	30.0	30.0
Universal gas constant	J/mol.K	8.314	8.314	8.314	8.314	8.314
Isentropic compressor work	J/mol	835.9	784.2	804.7	799.2	829.8
Isentropic efficiency	%	76.22	76.32	79.10	79.14	79.07
Actual compressor work	J/mol	1096.6	1027.4	1017.3	1009.8	1049.5
Mechanic efficiency	%	75.14	75.48	85.37	85.62	85.37
Motor efficiency	%	96.29	96.30	96.31	96.31	96.31
BHP	kW	343.0	321.9	361.8	360.8	371.7
$P_{comp,temp}$	kW	356.2	334.3	375.7	374.6	385.9

Table 4 Estimation value using the proposed model of the screw compressor in project B

Project B		Operating condition					
Parameter	Unit	On-peak block 1	Off-peak block 1	On-peak block 2	On-peak block 2	Off-peak block 2	Off-peak block 2
Mass flow rate	kg/h	19,555.2	15,447.6	18,274.0	18,273.6	16,268.4	16,268.4
Max mass flow rate	kg/h	21,000	21,000	22,000	22,000	22,000	22,000
Percentage mass flow rate	%	93.12	73.56	83.06	83.06	73.95	73.95
Suction pressure	barA	15.25	15.39	15.19	15.19	15.27	15.27
Discharge pressure	barA	29.120	28.810	29.020	28.861	28.952	28.810
Pressure ratio	–	1.91	1.87	1.91	1.90	1.90	1.89
Percentage head	%	88.35	85.48	88.09	87.08	87.15	86.24
Suction temperature	°C	29.10	28.82	29.99	29.60	30.24	29.90
Molecular weight	g/mol	17.6052	17.4941	18.747	18.7471	18.6581	18.6581
Specific heat ratio	–	1.3602	1.3610	1.3528	1.3528	1.3565	1.3565
Min suction pressure	barA	14.3	14.3	14.3	14.3	14.3	14.3
Max discharge pressure	barA	30.0	30.0	30.0	30.0	30.0	30.0
Universal gas constant	J/mol.K	8.314	8.314	8.314	8.314	8.314	8.314
Isentropic compressor work	J/Mol	1772.1	1711.8	1776.4	1757.8	1756.4	1739.8
Isentropic efficiency	%	79.86	79.06	79.61	79.61	79.09	79.09
Actual compressor work	J/mol	2219.1	2165.2	2231.3	2207.9	2220.8	2199.8
Mechanic efficiency	%	95.76	84.89	90.29	30.12	85.22	85.16
Motor efficiency	%	96.19	96.31	96.27	96.27	96.31	96.31
BHP	kW	715.0	625.6	669.2	663.4	631.2	625.6
$P_{comp,temp}$	kW	743.3	649.6	695.1	689.1	655.3	619.6

Table 5 Error of the proposed estimation methodology

Operating condition	Power consumption		
	Actual	Model	Error (%)
Project A			
Off-peak 16.05	366.19	356.17	−2.74
Off-peak 17.05	357.15	334.32	−6.39
Off-peak 15.40	389.85	375.70	−3.63
Off-peak 16.40	397.89	374.61	−5.85
Off-peak 17.40	398.83	385.90	−3.24
Project B			
On-peak block 1	777.45	743.29	−4.39
Off-peak block 1	648.15	649.56	0.22
On-peak block 2	700.26	695.14	−0.73
On-peak block 2	695.53	689.12	−0.92
Off-peak block 2	641.37	655.34	2.18
Off-peak block 2	634.11	649.61	2.45

power plants. After testing the model with specification data from the actual units, the result has illustrated that the proposed estimated model can accurately estimate power consumption in screw gas compressors under different operating conditions with an error of less than 5% when the compressor is operated at the maximum design condition. Thus, the estimation model can help reduce the time and simplify the process required to design a suitable screw gas compressor for natural gas power plant projects.

References

1. US energy information administration. International energy outlook (2016)
2. Odetayo, B., MacCormack, J., Rosehart, W.D., Zareipour, H., Seifi, A.R.: Integrated planning of natural gas and electric power systems. Int. J. Electr. Power Energy Syst. **103**, 593–602 (2018)
3. Mathur, S., Gosnell, G., Sovacool, B.K., Del Rio, D.D.F., Griffiths, S., Kim, M.B.J.: Industrial decarbonization via natural gas: a critical and systematic review of developments, socio-technical systems and policy options. Energy Res. Soc. Sci. **90**, 102638 (2022)
4. Kinnon, M.A.M., Brouwer, J., Samuelsen, S.: The role of natural gas and its infrastructure in mitigating greenhouse gas emissions, improving regional air quality, and renewable resource integration. Prog. Energy Combust. Sci. **64**, 62–92 (2018)
5. Miller, H., Nemec, T.: Mechanical Engineers' Handbook, Energy and Power, 3rd edn. Wiley, Hoboken (2006)
6. Bahadori, A.: Gas compressors. In: Natural Gas Processing, pp. 223–273. Gulf Professional (2014)
7. Heiyanthuduwage, M.A., Mounoury, S., Kovacevic, A.: Performance prediction methods for screw compressors. In: 7th International Conference on Compressors and their Systems 2011, Woodhead Publishing, pp. 411–420 (2011)

8. Wu, W., Zhang, Z., Peng, C., Li, X., Yang, Y., Lei, W.: Refreshed internal working characteristics of the single screw compressor based on experimental investigation. Int. J. Refrig. **143**, 118–125 (2022)
9. Li, Y., Xie, G., Sunden, B., Lu, Y., Wu, Y., Qin, J.: Performance study on a single-screw compressor for a portable natural gas liquefaction process. Energy. **148**, 1032–1045 (2018)
10. Ion, M., Valeriu, D., George, G.B., Ionut, P., Ştefan, Ş.A.: Numerical efficiency evaluation of a high pressure ratio screw compressor. In: 2018 5th International Conference on Mathematics and Computers in Sciences and Industry (MCSI), pp. 1–6 (2018)
11. Tomescu, S., Petrescu, V., Serban A., Voicu, S.: Energy efficiency of an oil injected screw compressor operating at various discharge pressures. In: 2021 10th International Conference on ENERGY and ENVIRONMENT (CIEM), pp. 1–5 (2021)
12. Brasz, J.J.J.: Comparison of part-load efficiency characteristics of screw and centrifugal compressors. In: International Compressor Engineering Conference, pp. 1–7 (2006)

An Approach for Voltage Drop Improvement in Distribution Line Using High-Voltage Capacitor Bank

Panapong Songsukthawan, Santipont Ananwattanaporn, Surakit Thongsuk, Natthanon Phannil, Sulee Bunjongjit, Chaichan Pothisarn, Suntiti Yoomak, Chaiyan Jettanasen, and Atthapol Ngaopitakkul

1 Introduction

The growth of industry and the socioeconomic development of a population result in an increase in energy consumption, which must be satisfied by electricity produced by power plants. The electric power system is operated and separated into four key sectors, namely, generation, distribution, distribution, and consumers, to deliver electrical energy to customers. As a result, the distribution systems must work harder to enhance their capacity for electric power transfer in order to keep up with the rising demand. The distribution systems, in contrast, are well known for their high R/X ratio, which causes a significant voltage drop throughout the distribution lines and causes significant power losses [1, 2]. Making the distribution lines efficient and effective is therefore crucial.

Voltage drop reduction has lately risen in importance in order to improve the electric power systems' dependability, efficiency, and service quality. Several

---

P. Songsukthawan
Center of Specialty Innovation, Bangkok University, Bangkok, Thailand

S. Ananwattanaporn · C. Pothisarn · S. Yoomak · C. Jettanasen · A. Ngaopitakkul (✉)
School of Engineering, King Mongkut's Institute of Technology Ladkrabang, Bangkok, Thailand
e-mail: atthapol.ng@kmitl.ac.th

S. Thongsuk
Faculty of Industrial Technology, Rajabhat Rajanagarindra University, Chachoengsao, Thailand

N. Phannil
Faculty of Engineering, Thaksin University (Phatthalung Campus), Phatthalung, Thailand

S. Bunjongjit
Faculty of Engineering, Rajamangala University of Technology Rattanakosin, Nakhon Pathom, Thailand

L. Chen (ed.), *Trends in Clean Energy Research*, Green Energy and Technology,
https://doi.org/10.1007/978-3-031-67987-2_11

approaches were motivated by the requirement for voltage drop improvement in electric power networks.

By managing dispersed energy resources in a balanced electrical power system, Paper [3] performed voltage regulation across the whole system using a created data-driven approach [4]. discusses an accurate and effective way for portraying voltage regulators based on both local and remote control modes in medium-voltage distribution systems. The bat algorithm and voltage stability index are used to choose the best size and position of the compensating devices in distribution systems in order to achieve voltage control and reduce power loss [5]. The ability to provide such higher voltage support makes battery energy storage technologies one of the most prospective flexibility providers in future power grids [6]. In order to provide voltage management and fulfill the necessary voltage profiles mandated by distribution system operators, article [7] provided the optimal coordination of any number of battery energy storage devices.

The voltage profile along a three-phase distribution line whose power transfer capability was increased by an uprating strategy based on the insertion of a third harmonic voltage component is explored in article [8] in order to enhance the capacity of the distribution line. In article [9], a new approach was developed to help the electric power system achieve the needed voltage stability margin by leveraging battery energy storage devices to provide both active and reactive power. The voltage source inverter (VSI), which injects the necessary missing voltage in series with the system voltage to rectify voltage fluctuations experienced in the lengths of the electric power system, is the central component of the dynamic voltage restorer (DVR) model for voltage profile enhancement in low-voltage distribution systems presented in Paper [10].

A multiverse optimizer (MVO) algorithm was used to show the best position and capacity for capacitor banks in radial distribution networks in [11]. The approach uses MATLAB's member and any commands to modify and partially use conventional loss sensitivity factors (LSFs) in order to reduce the search space for the optimization. Symbiotic organism search (SOS) and neural network algorithm (NNA) were integrated in article [12] to suggest the best design and operation of capacitor banks in radial distribution networks. The outcomes showed that yearly power loss mitigation is still being improved. A two-layer hybrid optimization method for the ideal placement, dimension, and switching of switched capacitors was reported in Paper [13]. The cost of generating active and reactive power in power plants, as well as active and reactive power losses in distribution networks, makes up the objective function. The findings demonstrate that the recommended method of capacitor bank placement and scheduled optimization has a stronger impact on voltage profile improvement. In order to reduce power loss, the placement and size of capacitor banks in distribution systems were analyzed using affine arithmetic in Paper [14]. In a publication, an optimal mathematical model was developed for the simultaneous reconfiguration of capacitor bank allocation in distribution systems [15]. The optimization problem was approached using a mixed-integer second-order cone programming model with the objective of minimizing overall investment in switchable and fixed capacitor banks as well as operating costs related to energy losses.

The vast majority of techniques that may be utilized to increase voltage drop, decrease power losses, and maintain voltage profile in order to increase the dependability, efficiency, and service quality of electric power systems were shown. The installation of capacitor banks has several benefits for electric power systems, including lower voltage drop, reactive power compensation, and energy losses. For this reason, capacitor bank use is exceptional in this work. For the purpose of resolving voltage drop issues, capacitor bank placement and capacity also need to be taken into consideration.

In this article, we use PSCAD software to examine the variables that have an impact on the voltage drop on a 22-kV overhead distribution line in Thailand, both with and without the installation of capacitor banks. The variables that affect distribution line length, electric load power factor, and electric load capacity are modified, and the transmitting and receiving sides of the distribution line are used to examine the power quality metrics of apparent power, actual power, reactive power, current, and voltage. Additionally, a mathematical solution for the study of voltage drops from capacitor banks mounted on distribution lines is supplied and contrasted using capacitor bank capacities and locations.

2 Case Study Distribution System

The impact of voltage drop in the distribution line of Thailand was investigated using PSCAD software. The 22-kV overhead distribution line with π-network comprising resistance, inductance, and capacitance was selected and modeled for study cases. The simplified diagram and PSCAD diagram of the distribution system in case of without and with high-voltage capacitor placement can be demonstrated as shown in Fig. 1. From the figure, it can be seen that the 22-kV distribution system consists of substation represented by the three-phase alternating current source (position 1) supply power to the 2.5 MVA load (position 5) by using the distribution line (position 3). The power factor of the load is 0.85, which is the minimum power factor specified by the Provincial Electricity Authority. The measurement point on both substation and load side were installed (positions 2 and 4) to obtain the system voltage, current, actual power, and reactive power.

3 Proposed Capacitor Placement

From the simulation without installation of capacitor bank, the result shows that overhead power distribution lines at the 22-kV voltage level will have a voltage drop close to 5% every time, with a distance of 30 kilometers. Therefore, the placement of high-voltage capacitor bank on the distribution line consists of a two-capacitor

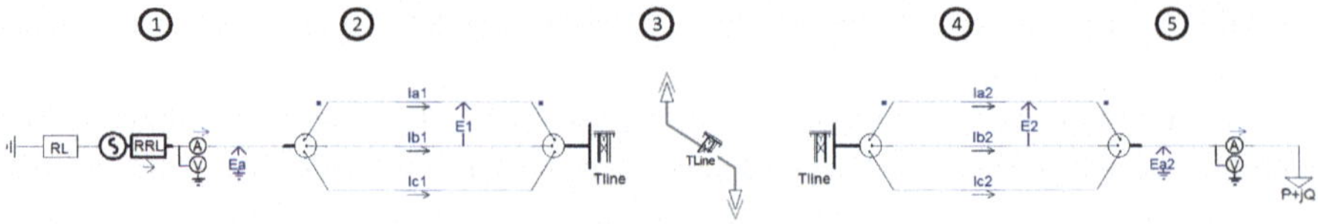

(a) PSCAD diagram

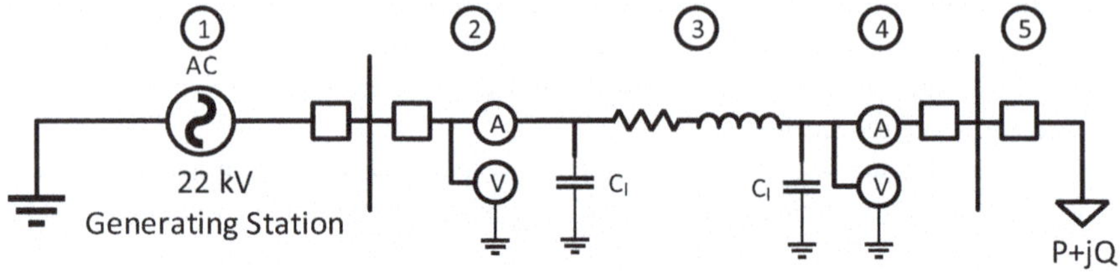

(b) Simplified diagram

Fig. 1 22-kV distribution line based on Thailand's PEA distribution system

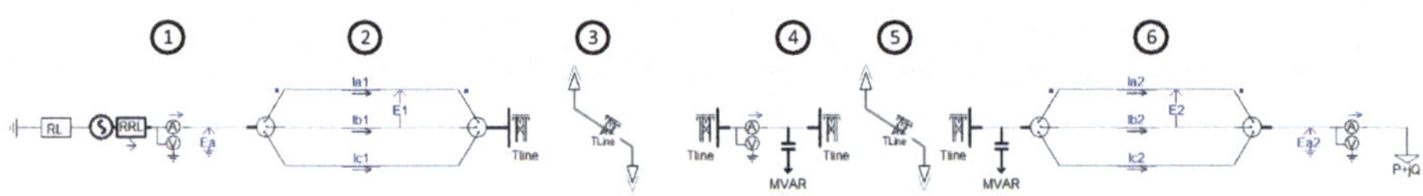

(a) PSCAD diagram

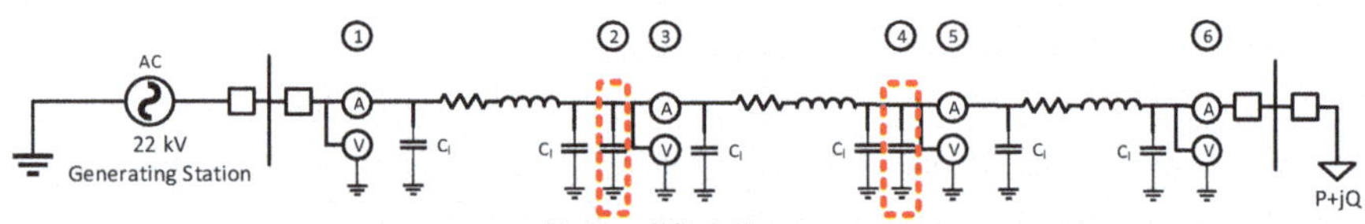

(b) Simplified diagram

Fig. 2 The proposed placement of capacitor bank on distribution line

unit with various capacity sizing. Two capacitors are installed at the same distance between positions 2 and 4 as shown in Fig. 2. The size of the capacitor used in each installation location is calculated from the size of the desired capacitor divided by 2, and the power meter is used to collect the voltage, actual power, and reactive power.

4 Result

The voltage and current from the sending and receiving sides for the cases with and without the capacitor bank installation are presented in Fig. 3a, b, respectively. In Fig. 3, the distribution line distance is 80 km and the electric load capacity is set to 2.5 MVA with a power factor of 0.85. The result obtained from the measurement unit

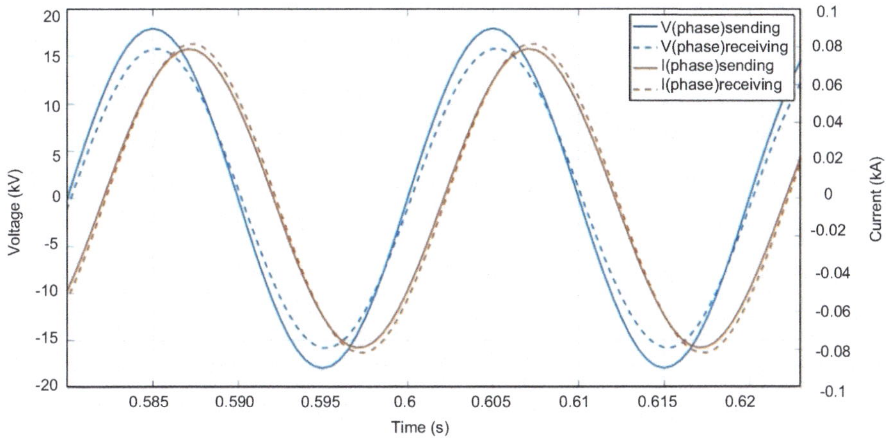

(a) Without capacitor bank installation

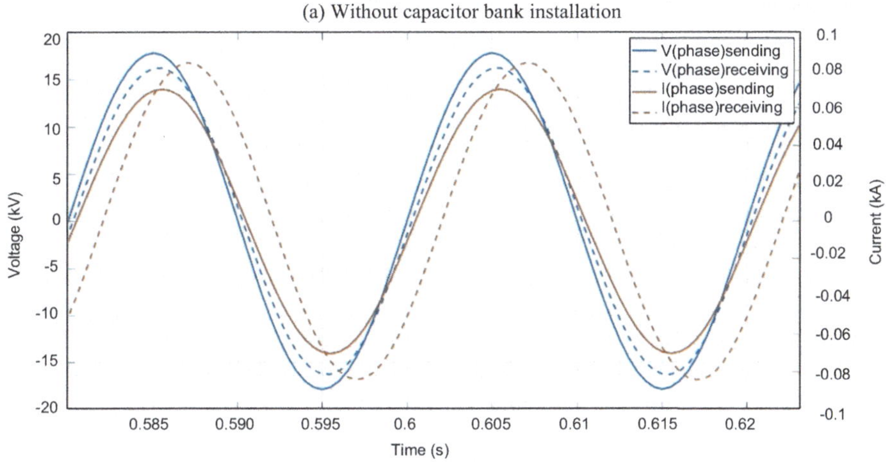

(b) With capacitor bank installation

Fig. 3 Comparison of current and voltage waveforms from sending and receiving sides

is summarized in Table. 1. In the case of without capacitor installation, the current waveform of the sending side leads to the voltage waveform, since the reactance occurred in the distribution line, while the difference of the voltage and current waveforms at the receiving side is dependable on the electric load power factor. For capacitor bank installation, the difference of the angles of the current and voltage waveforms are decreased, as the reactance on the distribution line is compensated by the reactance from the capacitor, resulting in less reactive current loss. In addition, the capacitor bank raises the receiving side voltage, increasing the current supplied to the electrical load. However, the phase difference between the current and voltage of the receiving sides is the same as the case without capacitor bank installation, as the electric power factor is not changed.

Table 1 Results in terms of voltage level from the proposed capacitor placement

Total capacitor size (MVAr)		0.8	0.9	1	
1/2 of total capacitor size (MVAr/ 1 unit)		0.4	0.45	0.5	Without capacitor
Real power (MW)	*Sending*	1.8439	1.8532	1.8628	1.7790
	Receiving	1.7334	1.7435	1.7536	1.6556
Reactive power (MVAr)	*Sending*	0.4659	0.3759	0.2856	1.1710
	Receiving	1.0744	1.0806	1.0868	1.0261
Voltage at W (kVLL, rms)		21.9011	21.9016	21.9020	21.8974
Voltage at X (kVLL, rms)		21.2911	21.3317	21.3724	20.9719
Voltage at Y (kVLL, rms)		20.5063	20.5656	20.6252	20.0407
Voltage at Z (kVLL, rms)		19.8691	19.9266	19.9843	19.4180
% Voltage drop		10.2269	9.9111	9.5958	12.7683

5 Conclusion

This study examines the causes of voltage drops and distribution line enhancements made possible by capacitor banks. The voltage drop is decreased by using a bigger capacitor bank. As a result, the amount of total capacitor bank capacity used has a big impact on the voltage drop. Thus, the suggested strategy reduces voltage loss. To enhance voltage along the whole distribution line, the position of the capacitor bank must be taken into account. By positioning capacitor banks far from the transmitting side or the location with the greatest voltage drop, the voltage drop issue can be made worse. As a result, the need for voltage support dictates the proper position for voltage support. In the future work, the optimization of capacitor bank location and capacity on the distribution line for voltage drop improvement is conducted based on using the proposed mathematical equation.

References

1. Elsayed, A.M., Hegab, M.M., Farrag, S.M.: Smart residential load management technique for distribution systems' performance enhancement and consumers' satisfaction achievement. Int. Trans. Electr. Energy Syst. **29**(4), 1–23 (2019)
2. Devabalaji, K.R., Yuvaraj, T., Ravi, K.: An efficient method for solving the optimal sitting and sizing problem of capacitor banks based on cuckoo search algorithm. Ain Shams Eng. J. **9**(4), 589–597 (2018)
3. Xu, H., Domínguez-García, A.D., Veeravalli, V.V., Sauer, P.W.: Data-driven voltage regulation in radial power distribution systems. IEEE Trans. Power Syst. **35**(3), 2133–2143 (2020)
4. Nasef, A., Shaheen, A., Khattab, H.: Local and remote control of automatic voltage regulators in distribution networks with different variations and uncertainties: practical cases study. Electr. Power Syst. Res. **205**, 107773 (2022)
5. Yuvaraj, T., Devabalaji, K.R., Srinivasan, S., Prabaharan, N., Hariharan, R., Alhelou, H.H., Ashokkumar, B.: Comparative analysis of various compensating devices in energy trading radial distribution system for voltage regulation and loss mitigation using Blockchain technology and bat algorithm. Energy Rep. **7**, 8312–8321 (2021)

6. Xie, Y., Liu, L., Wu, Q., Zhou, Q.: Robust model predictive control based voltage regulation method for a distribution system with renewable energy sources and energy storage systems. Int. J. Electr. Power Energy Syst. **118**, 105749 (2020)
7. Zhao, T., Parisio, A., Milanović, J.V.: Distributed control of battery energy storage systems in distribution networks for voltage regulation at distribution–distribution network interconnection points. Control. Eng. Pract. **119**, 104988 (2022)
8. Zhang, L., Shen, C., Chen, Y., Huang, S., Tang, W.: Coordinated allocation of distributed generation, capacitor banks and soft open points in active distribution networks considering dispatching results. Appl. Energy. **231**, 1122–1131 (2018)
9. Pereira, L.D.L., Yahyaoui, I., Fiorotti, R., de Menezes, L.S., Fardin, J.F., Rocha, H.R.O., Tadeo, F.: Optimal allocation of distributed generation and capacitor banks using probabilistic generation models with correlations. Appl. Energy. **307**, 118097 (2022)
10. Almabsout, E.A., El-Sehiemy, R.A., An, O.N.U., Bayat, O.: A hybrid local search-genetic algorithm for simultaneous placement of DG units and shunt capacitors in radial distribution systems. IEEE Access. **8**, 54465–54481 (2020)
11. Mtonga, T.P.M., Kaberere, K.K., Irungu, G.K.: Optimal shunt capacitors' placement and sizing in radial distribution systems using multiverse optimizer. IEEE Can. J. Electr. Comput. Eng. **44**(1), 10–21 (2021)
12. Nguyen, T.P., Nguyen, T.A., Phan, T.V.H., Vo, D.N.: A comprehensive analysis for multi-objective distributed generations and capacitor banks placement in radial distribution networks using hybrid neural network algorithm. Knowl.-Based Syst. **231**, 107387 (2021)
13. Jafari, A., Ganjehlou, H.G., Khalili, T., Mohammadi-Ivatloo, B., Bidram, A., Siano, P.: A two-loop hybrid method for optimal placement and scheduling of switched capacitors in distribution networks. IEEE Access. **8**, 38892–38906 (2020)
14. Adusumilli, B.S., Raj, V., Adusumilli, V.B.: Capacitor placement and sizing to minimize losses in a radial distribution network considering uncertainty using modified affine arithmetic division. Sustain. Energy Grids Netw. **27**, 100492 (2021)
15. Home-Ortiz, J.M., Vargas, R., Macedo, L.H., Romero, R.: Joint reconfiguration of feeders and allocation of capacitor banks in radial distribution systems considering voltage-dependent models. Int. J. Electr. Power Energy Syst. **107**, 298–310 (2019)

Numerical Analysis of the Flow Phenomena Inside the Vortex Tube with Different Turbulence Models

Osama Ali Ahmed Awan, Robert Sager, Thomas Gier, Manfred Wirsum, and Ekachai Juntasaro

1 Introduction

Energy is the backbone for urbanization and technological advancement of the world. However, the current challenge for the world is to produce sustainable energy to address environmental concerns. At the same time, there is a need for greater efficiency and a reduction in energy waste. This can be achieved by inclusion of new devices into existing systems. One such device is the vortex tube, which is a simple thermal device with no moving parts. The first idea of a vortex tube was given by the French scientist Georges J. Ranque in 1933. More than 10 years later, in 1947, the German physicist Rudolf Hilsch gave an adequate explanation of the temperature separation effect (TSE) inside the vortex tube. For this reason, the vortex tube is commonly referred to as the Ranque-Hilsch vortex tube (RHVT) [1]. The RHVT is an expansion device that separates the incoming high-pressure stream into hot and cold streams. The resulting hot stream has a higher temperature than the incoming high-pressure stream, while the subsequent cold stream has a lower temperature than the incoming high-pressure stream. Although the RHVT holds great potential for different applications, it has certain limitations such as suboptimal efficiency

O. A. A. Awan (✉)
The Sirindhorn International Thai-German Graduate School of Engineering, KMUTNB, Bangkok, Thailand

Institute of Power Plant Technology, Steam and Gas Turbines, RWTH Aachen University, Aachen, Germany

R. Sager · T. Gier · M. Wirsum
Institute of Power Plant Technology, Steam and Gas Turbines, RWTH Aachen University, Aachen, Germany

E. Juntasaro
The Sirindhorn International Thai-German Graduate School of Engineering, KMUTNB, Bangkok, Thailand

L. Chen (ed.), *Trends in Clean Energy Research*, Green Energy and Technology,
https://doi.org/10.1007/978-3-031-67987-2_12

and limited development. Therefore, much research is still being carried out to find the optimum geometric and operating parameters to exploit the advantages of the RHVT.

Primarily, there are three different types of research in the field of vortex tubes: experimental, numerical, and analytical. Computational research has the advantages of flexibility, cost-effectiveness, time-efficiency, and adaptability over the experimental and analytical approaches. Therefore, akin to other fields, more emphasis has been put on computational research of the RHVT. While several researchers have explored the interesting advances and flow phenomena within the RHVT, the answer to one question remains vague: Which is the best turbulence model for the numerical analysis of the RHVT? Different authors have different opinions on this, which makes it difficult to answer. Table 1 shows the turbulence models considered and the best performing ones by various researchers in the recent past. The choice of a suitable turbulence model is of significant importance for the accurate numerical prediction of the flow within the RHVT. Therefore, the authors of this paper have attempted to address this issue by considering standard k-ε (k-ε), standard k-ω (k-ω), shear stress transport k-ω (SST k-ω), and renormalization group k-ε (RNG k-ε) in the current study. Several additional turbulence models, namely, scale adaptive simulation shear stress transport (SAS SST), Spalart-Allmaras (SA), Reynolds stress equation model (RSM), and realizable k-ε, have been considered by various authors. However, these were not included in the current study as they did not perform well based on the available literature.

The primary objective of this work is to determine which turbulence model produces the most accurate TSE when compared with reference data. This is accomplished through numerical simulations conducted on a fully three-dimensional model of the RHVT. The details of this research study unfold across the following sections of this paper. Section 2 outlines the geometry under consideration and the reference geometry. It also discusses how to obtain a suitable mesh. Section 3 presents the results of the research. Finally, the last section encapsulates the conclusions and outlines the future objectives of this research.

2 Numerical Study

2.1 Geometry

In some of the previous studies, a 2D geometric model was considered because its result had a close qualitative resemblance to the 3D analysis results [4]. However, quantitively, it lacks accuracy to predict accurate results. Therefore, a 3D model was selected in this study. The geometric model of the considered RHVT model with dimensions is shown in Fig. 1. The current geometry is very similar to the geometry of Rafiee and Sadeghiazad [9] and Kaya et al. [10]. Table 2 shows the comparison between the present geometry and the reference geometry.

Table 1 Different turbulence models used in recent research studies only

Research paper	Investigated turbulence models								Best turbulence model
	k-ε	k-ω	SST k-ω	SAS SST	SA	RNG k-ε	RSM	Realizable k-ε	
Singh et al. [2]	✓	✓	✓	X	X	✓	X	X	k-ε
Dyck et al. [3]	✓	✓	✓	✓	X	X	X	X	SAS SST
Lagrandeur et al. [4]	✓	X	✓	X	X	X	X	X	k-ω SST
Sadeghiseraji et al. [5]	✓	✓	✓	X	X	X	X	X	k-ε
Hu et al. [6]	✓	X	✓	X	✓	X	X	✓	Realizable k-ε
Bazgir et al. [7]	✓	X	✓	X	X	✓	X	✓	RNG k-ε
Chýlek et al. [8]	✓	✓	✓	X	X	X	✓	✓	k-ε

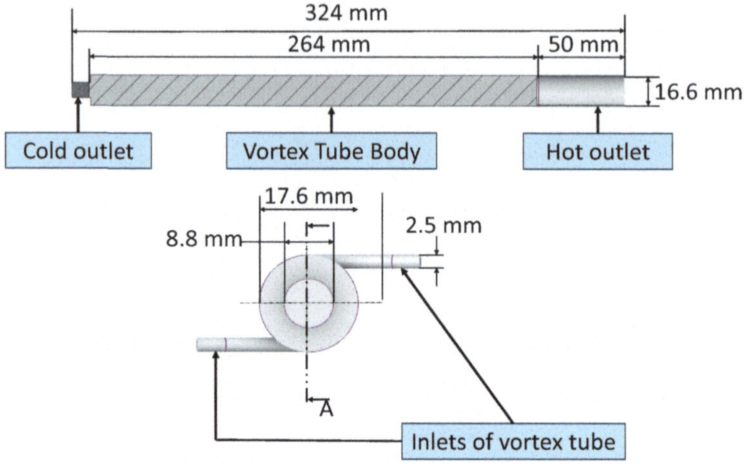

Fig. 1 Dimensions of the vortex tube model considered in this paper

Table 2 Comparison of geometric parameters of present model with reference models

RHVT parameter				
Name	Unit	This study	Rafiee and Sadeghiazad [9]	Kaya et al. [10]
Length	mm	264	250	100
Diameter	mm	17.6	18	7
L/D	N/A	15	13.89	14.28
Nozzle number	N/A	2	2	2
Nozzle diameter	mm	2.5	1.7	3.38

2.2 *Mesh Study*

ANSYS 2022 R1 was used for the current numerical study. A 3D CFD mesh was generated on ANSYS meshing. The mesh consists of tetrahedral elements with a size of 0.75 mm. Inflation layers are augmented along the wall to refine the spatial resolution near the boundary walls. The accurate representation of these boundary layers is based on the correct incorporation of these inflation layers. This type of meshing results in robustness of the computational study and leads to more accurate simulations of the problem. As a next step, a grid-independent study was carried out, which is very important to understand how the simulation results change when the grid or mesh resolution is changed. This has a direct influence on the numerical accuracy and optimization of the computational effort. Therefore, the focus of the mesh study in this paper lies on refining the mesh at the inlet and outlet regions of the vortex tube until consistent results are obtained. The sphere of influence function of ANSYS meshing is used to specify the element size. The mesh refinement is carried out until the results remain unchanged. The final element sizes at the inlet, cold outlet, and hot outlet are 0.288 mm, 0.35 mm, and 0.4 mm, respectively. The

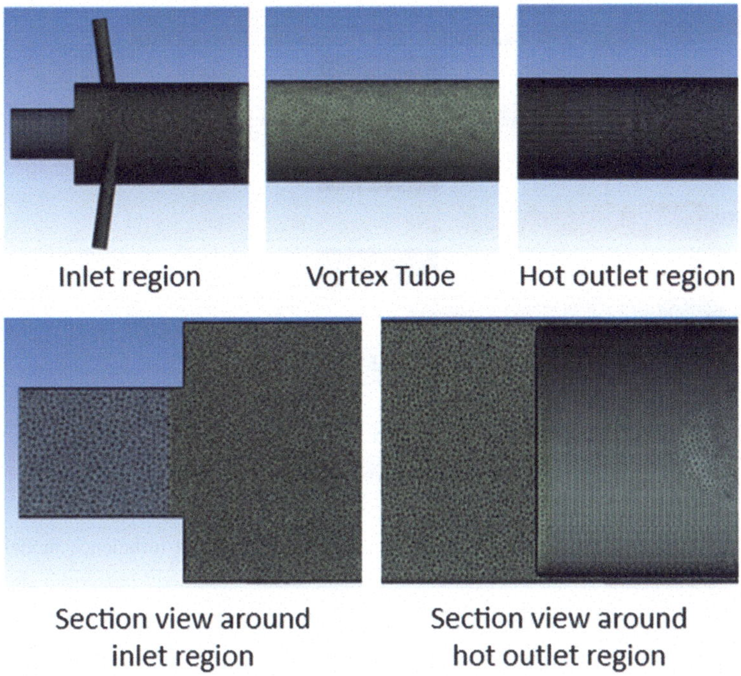

Fig. 2 Regions of mesh refinement and mesh at different regions of the vortex tube

selected mesh has 8.19 million elements and 2.65 million nodes. Figure 2 shows the mesh of the vortex tube model used in this study. When considering the low Reynolds number (low Re) turbulence models, the y-plus (y^+) value plays a very crucial aspect of the numerical study. It is the nondimensional distance of the first cell from the wall, and it defines the quality of the mesh in the boundary layer. A y^+ study refers to changing the first layer height of the cell near the wall until a required y^+ value is obtained. In this work, most parts of the RHVT have a y^+ of 1 because this is required for low-Re models. However, the choice of turbulence model does not depend on the y^+ value of the mesh, and we do not want to confuse the discussion of the y^+ value with the choice of turbulence model. Therefore, the same mesh is used to evaluate all turbulence models.

2.3 *Boundary Conditions and Numerical Solution*

The hot and cold outlets are pressure-type outlets with a value of 1.01 bar, while at the inlet, a relative pressure of 2.5 bar is specified. The static temperature at the inlet is set to 294.2 K. Adiabatic no-slip conditions are applied to the inner walls of the vortex tube. These values are the same as those used in the research work of Rafiee

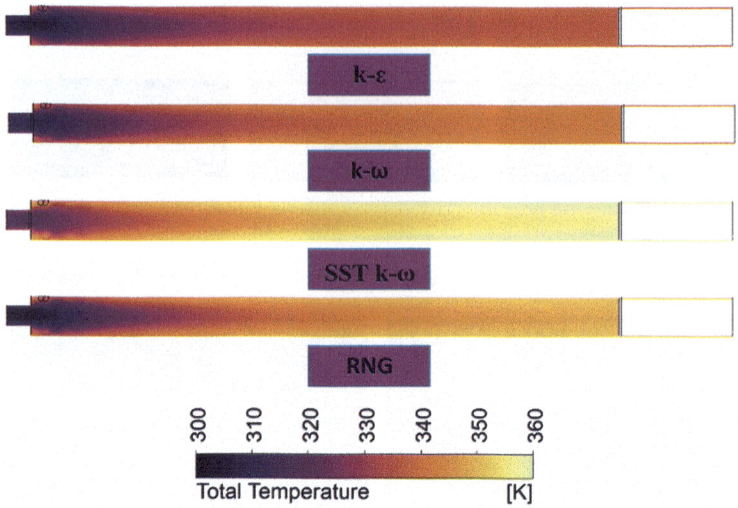

Fig. 3 The obtained local contours of total temperature by different turbulence models at the midplane of the RHVT

and Sadeghiazad [9]. Furthermore, the steady-state simulations are carried out on ANSYS CFX. Air as an ideal gas is chosen as the working fluid. The convergence limit for the residuals is set to 1e-6. In this study, k-ε, k-ω, SST k-ω, and RNG k-ε are considered based on the results of previous research studies as indicated in Table 1.

3 Results and Discussion

This section discusses the results obtained from different turbulence models. The experimental results of Rafiee and Sadeghiazad [9] are used to validate the results of this study. The present comparison is mainly based on the temperature distribution inside the RHVT. Figure 3 shows the total temperature contours, at the midplane of the vortex tube, obtained by different turbulence models. It was observed that all turbulence models were able to generate hot and cold sections inside the RHVT. It can be concluded from the results that the choice of turbulence model does not qualitatively affect temperature separation. In addition, it may be noted that the regions closer to the wall have higher temperatures than the region at the core of the RHVT. Moreover, it may be noted that the values of the maximum and minimum temperatures are substantially influenced by the choice of the turbulence model. If the local contours of the total temperature, at the midplane, are qualitatively compared with the experimental results of Rafiee and Sadeghiazad [9], the k-ε and SST k-ω turbulence models have the closest resemblance.

Table 3 Deviation of results present from reference results

	Turbulence model			
	k-ε	k-ω	SST k-ω	RNG k-ε
Average % (ΔT_c)	7.18	7.87	22.74	35.7

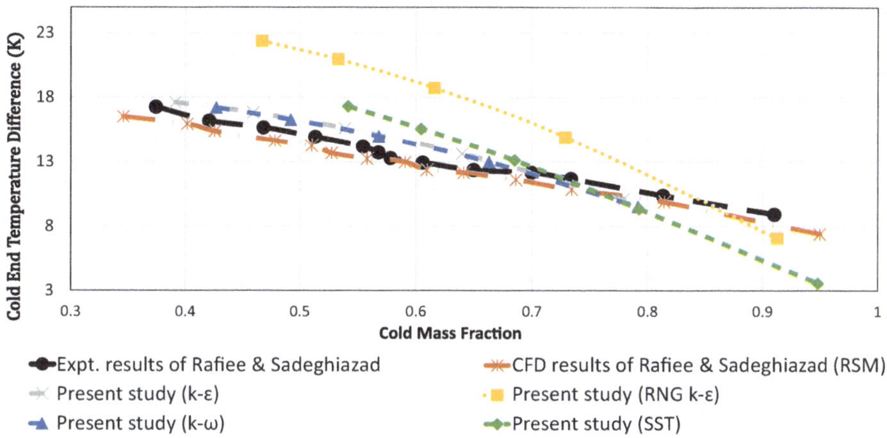

Fig. 4 Comparison of results at different cold mass fractions

The obtained results are compared with the reference data to find out which turbulence model predicted results closest to the reference data. To compare the results obtained from different turbulence models, the percentage deviation of the cold outlet temperature difference (ΔT_c) of numerical results from experimental results is used. Table 3 shows the average deviation of the results obtained by different turbulence models from that of reference results. Figure 4 represents the comparison of the experimental and computational results of Rafiee and Sadeghiazad [9] with the results of the present study.

It can be seen that the results of the k-ε turbulence model have close resemblance with the reference results. This is followed by the k-ω turbulence model that also shows good agreement with the reference results. However, there is a significant difference in the results of the other two models when compared with reference results. The close resemblance of the results of the k-ε model to the reference data supports the conclusions drawn in previous studies on the same topic, as demonstrated in Table 1.

In addition, Rafiee and Sadeghiazad [9] have used the Reynolds stress model (RSM) model in their study. They were able to capture almost similar results to the experimental results. However, the RSM model is expensive in terms of computational cost. Nevertheless, if the computation cost allows, then the RSM model might be a more suitable choice for numerical investigations of the RHVT. Furthermore, the hot end temperature difference is also an important parameter to evaluate which turbulence model is the best. However, Rafiee and Sadeghiazad [9] have only considered the cold end temperature difference to evaluate the performance of the

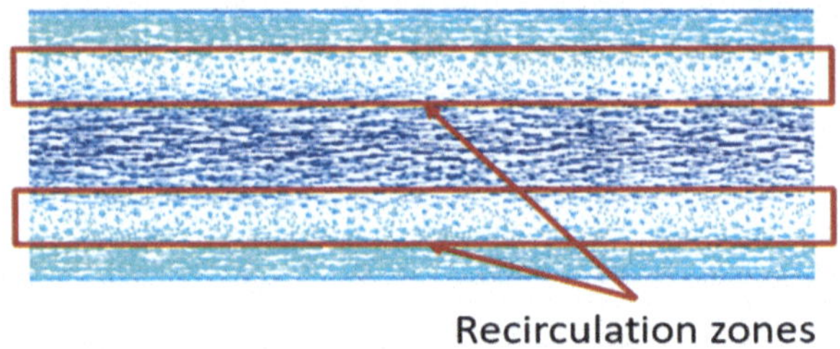

Fig. 5 Recirculation zones visible in the plot of velocity vectors at the midplane of the RHVT

vortex tube. This is why only the cold end temperature differences are available for validation. Nevertheless, we will consider the hot end temperature difference as an evaluation parameter in our subsequent numerical investigations.

Moreover, when the velocity vectors were analyzed from the numerical results, a recirculation zone was identified between the hot and cold flows inside the RHVT. Figure 5 visually depicts this intriguing recirculation zone. To enhance the accuracy of the predicted results, it is proposed to refine the mesh around this area by dividing the RHVT into a pair of concentric cylinders. This will be investigated in our next numerical investigations of the RHVT.

4 Conclusions

In this study, a 3D numerical analysis was performed to find the best turbulence model that accurately reflects the real vortex tube behavior. Four different turbulence models are used for the investigation, namely, k-ε, k-ω, SST k-ω, and RNG k-ε. The numerical results presented in this study are validated and compared with the experimental data of Rafiee and Sadeghiazad [9]. The geometry in this study and the geometry of Rafiee and Sadeghiazad [9] are almost the same; however, there are differences between them. In terms of the results, all turbulence models were able to qualitatively capture the temperature separation effect. However, the magnitudes of cold and hot temperatures were different for each turbulence model. No turbulence model was able to predict the exact temperature magnitudes when compared to the reference data. The results of k-ε and SST k-ω models were close to the reference data when considering the total temperature contours. However, the k-ε model exhibits the best agreement with the reference data when considering the cold exit temperature difference at different cold mass fractions. The average difference between the results for the k-ε turbulence model is 7.18%. However, it is worth noting that the geometry of the RHVT also exhibits a small difference, which could be a contributing factor to the variation in results. Nevertheless, the obtained results for the k-ε turbulence model are almost consistent with the conclusions of previous studies. Hence, it will be used for our future numerical studies. Moreover, a recirculation zone was identified between the hot and cold flows inside the RHVT.

The mesh around the recirculation zone will be further refined in the future. This may lead to a more accurate prediction of results.

References

1. Ambedkar, P., Dutta, T.: Analysis of various separation characteristics of Ranque-Hilsch vortex tube and its applications-a review. Chem. Eng. Res. Des. (2023)
2. Singh, R.K., Pramanick, A.K., Rana, S.C.: Computational study of temperature separation for a three-dimensional vortex tube with cold exit diameter and nozzle number variation. Int. J. Ambient Energy. **43**, 7046–7060 (2022)
3. Dyck, N.J., Parker, M.J., Straatman, A.G.: The impact of boundary treatment and turbulence model on CFD simulations of the Ranque-Hilsch vortex tube. Int. J. Refrig. **141**, 158–172 (2022)
4. Lagrandeur, J., Croquer, S., Poncet, S., Sorin, M.: A 2D numerical benchmark of an air Ranque-Hilsch vortex tube based on a fractional factorial design. Int. Commun. Heat Mass Transfer. **125**, 105310 (2021)
5. Sadeghiseraji, J., Moradicheghamahi, J., Sedaghatkish, A.: Investigation of a vortex tube using three different RANS-based turbulence models. J. Therm. Anal. Calorim. **143**, 4039–4056 (2021)
6. Hu, Z., Li, R., Yang, X., Yang, M., Zhang, Y.: Numerical simulation for three-dimensional flow in a vortex tube with different turbulence models. Numer. Heat Transf. A Appl. **77**, 121–133 (2020)
7. Bazgir, A., Khosravi-Nikou, M., Heydari, A.: Numerical CFD analysis and experimental investigation of the geometric performance parameter influences on the counter-flow Ranque-Hilsch vortex tube (C-RHVT) by using optimized turbulence model. Heat Mass Transf. **55**, 2559–2591 (2019)
8. Vortex Tube: A Comparison of Experimental and CFD Analysis Featuring Different RANS Models, vol. EDP Sciences (2018)
9. Rafiee, S.E., Sadeghiazad, M.M.: Three-dimensional and experimental investigation on the effect of cone length of throttle valve on thermal performance of a vortex tube using k–ε turbulence model. Appl. Therm. Eng. **66**, 65–74 (2014)
10. Kaya, H., Günver, F., Kirmaci, V.: Experimental investigation of thermal performance of parallel connected vortex tubes with various nozzle materials. Appl. Therm. Eng. **136**, 287–292 (2018)

Technical-Economic Implementation of a Control and Supervision System in Air-Conditioning Systems Through BACnet for Energy Efficiency in Buildings

Gary Ampuño, Jose Jaime-Carriel, Pedro Nunez-Izaguirre, and Juan Lata-García

1 Introduction

Given the current climatic situation of the planet and other independent factors, it is common to find the use of air-conditioning equipment in most buildings to achieve more comfortable thermal sensations. It is possible to find several models of air-conditioning equipment which are controlled manually, but this does not generate an optimal control of the air-conditioning increasing the consumption of electrical energy. Studies such as those conducted by [1–4] developed tests in different types of buildings in a coastal climate zone showing that energy and gas consumption are increased by the types of control and even by the architecture and type of materials of the buildings.

Strategies such as monitoring systems improve consumption conditions because they involve a process of taking values over a specific period; such real-time taking of values is systematized and is responsible for measuring the progress and results of a set of activities, through observation and collection of information in each of the processes, to ensure results. In addition, it allows the direct involvement of the system's beneficiaries. Projects are implemented in the CIESOL research center in Spain [5, 6] on optimal and multiobjective control to reduce the consumption of areas with thermal comfort for users. Here, techniques such as solar collectors are also used to reduce consumption.

This paper covers a study and implementation of the data communication protocol BACnet (Building Automation and Control Networks) which is a data communication protocol for Building Automation and Control Networks used in building automation and control networks as an alternative for the consumption of air conditioners in a higher education institution. In works such as [7–9], results

G. Ampuño (✉) · J. Jaime-Carriel · P. Nunez-Izaguirre · J. Lata-García
Universidad Politécnica Salesiana, Cuenca, Ecuador
e-mail: gampuno@ups.edu.ec

L. Chen (ed.), *Trends in Clean Energy Research*, Green Energy and Technology,
https://doi.org/10.1007/978-3-031-67987-2_13

are shown in the comfort of buildings and industries; this due to the optimal use of measurement variables with the use of BACnet in HVAC that are responsible for the air-conditioning of buildings.

The results presented by the researchers are given in an office that works with attention to students 8 h a day, and you can see how energy consumption decreases up to 50% by the correct control techniques applied from the BACnet incorporating an old air-conditioning equipment. For the test, a BACnet module was adapted to the air-conditioning equipment, and a network analyzer was used to collect consumption data from the equipment.

2 Climate Control System

2.1 Scheme of the Control System

For the tests performed, Fig. 1 shows the scheme of the control system that incorporates the BACnet system and the connection of temperature sensors and thermostats (daisy chain connection). Later, with the acquired data, a contrast of the consumptions on different days is made.

The BACnet protocol is a communication protocol generally used in the integration of building automation systems. Among the systems that are integrated in these automations are access control, fire systems, lighting, and automatic air-conditioning systems. The BACnet achieves communication between supervisory devices and field devices. It is an open standard that seeks the integration of devices regardless of brands [10, 11]. The daisy chain connection is a successive connection of devices, i.e., reaches with the physical bus to the communication port of a device and from this port goes to the next device and so on. In this way, we can link the number of devices that allows us our main supervisor depending on the capacity of this [12–14].

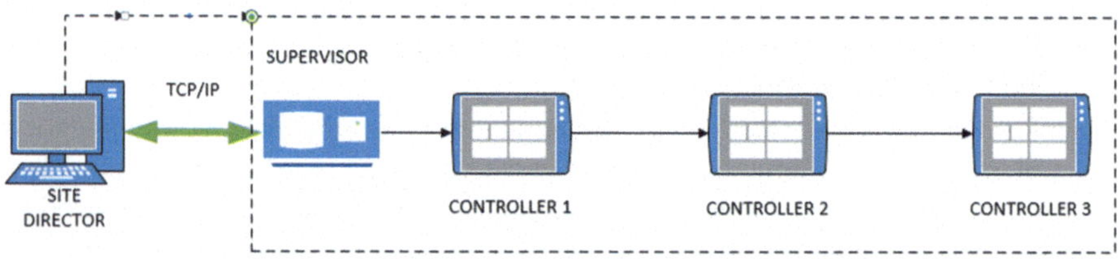

Fig. 1 Proposed network implemented

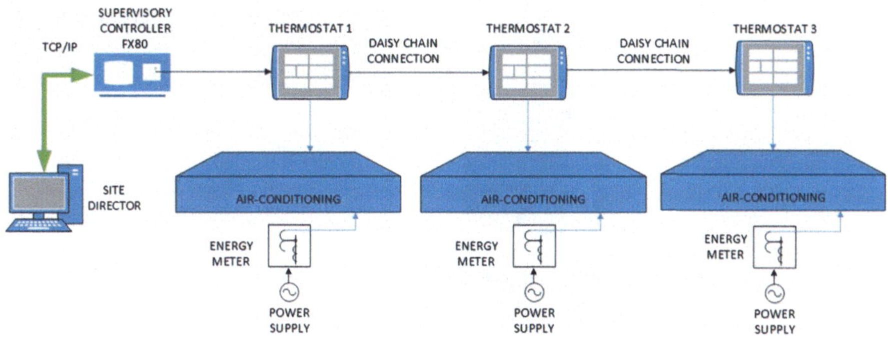

Fig. 2 Power and control connections in air-conditioning equipment

2.2 *Power and Control Connections in Air-Conditioning Equipment*

The present work implemented a practical module with which it is possible to remotely monitor and control (SCADA) three air-conditioning units, two of which are in modules and a floor-ceiling air-conditioning unit in the laboratory. For the analysis of energy consumption, Fig. 2 shows three air-conditioning units that transmit their temperature values (coming from the thermostats) to the controller by means of the daisy chain topology. A network analyzer was connected to the power supply of the air-conditioning units to measure the power consumed in the tests, as shown in Fig. 3. To achieve the communication network, a supervisor unit was used that works through a BACnet and TCP/IP connection. The BACnet communication foresees connecting the field devices (thermostat), which have a network card that allows them to communicate with the master or supervisor, in order to visualize the air-conditioning system from a laptop or PC. This was done through TCP/IP communication. Once the communication between all the devices was done, the programming was done from the supervisor's software through the laptop.

We perform tests with the normal operation (without the control system) of the equipment in order to obtain and later incorporate the BACnet module with the controller to compare the energy consumption and thus obtain energy savings by mitigating the misuse of the equipment. All the information is displayed through a SCADA system created in the application that shows in real time the data of the equipment connected to the network.

2.3 *Module BACnet: PID Controller*

As will be explained below, since the model of the air-conditioning system is expressed as a closed-loop system, some of the inputs must be delayed to reproduce

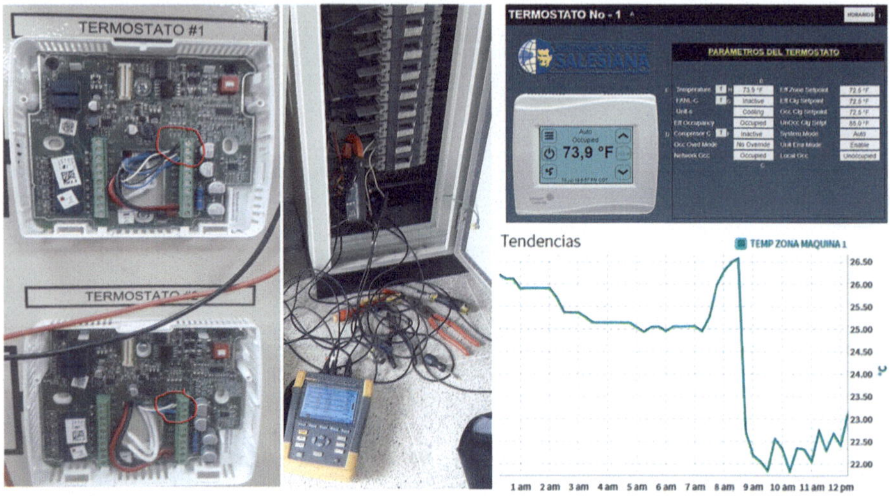

Fig. 3 Test preparation. (**a**) Daisy chain connection of the thermostats, (**b**) connection of the Znetwork analyzer for data collection of the consumed power, and (**c**) SCADA system with the options of monitoring and trending of temperatures in real time

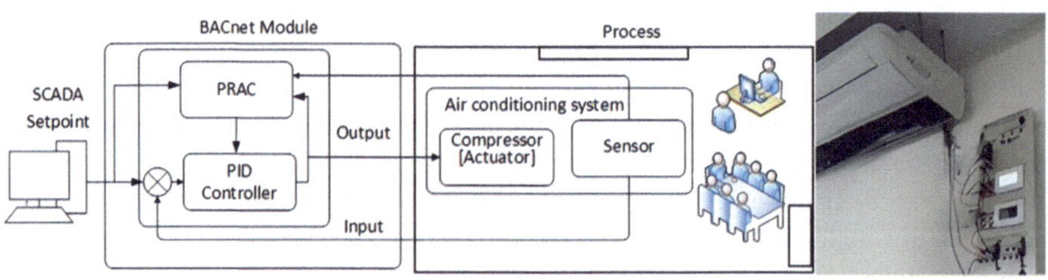

Fig. 4 Control system for implementing PID + PRAC

the dynamic behavior of the real plant. Fig. 4 presents the control model that governs the air-conditioning system, where the setpoint is entered through the SCADA system, which in turn allows to see in real time the temperatures in the applied area. The measurements are made by means of the sensor incorporated in the air-conditioning system. By programming the software Metasys System PID + PRAC of the manufacturer Johnson Controls, the error obtained between the subtraction of the setpoint is entered, and the actual value of the temperature to enter the controller model delivers the control signal to the actuator.

The PID + PRAC controller belongs to the manufacturer Johnson Controls and is the sum of a proportional, integral, and derivative controller plus the algorithm pattern recognition adaptive control (PRAC) [15, 16]. This PID + PRAC set makes the controller regulate its management over time adapting to the environment and improving the energy consumption of the actuator. Figure 5 shows a partial

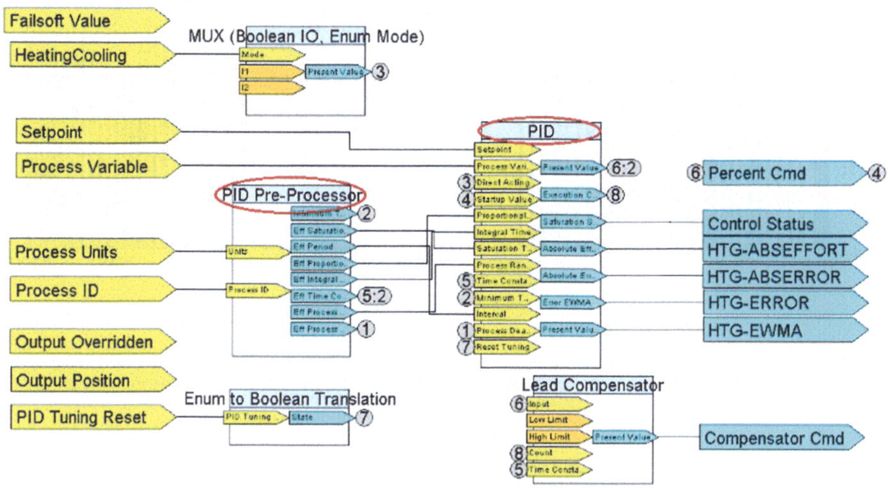

Fig. 5 PID + PRAC control programming block

presentation of the implemented software controller block for a better understanding of the software tool used.

3 Comparison Between Experimental Data

This section shows the results of the models described above and compares them with real data from the installation. To carry out this comparison, two different types of experiments have been performed. To study the efficiency of the air-conditioning system, it is basic to keep the temperature set at a comfort temperature using only the thermostat of the air-conditioning system. In the other way of working, the BACnet module is incorporated to be in charge of the temperature collection and control using the SCADA and the previously programmed controller. To perform the simulations, the configurations presented above have been connected as shown in Fig. 3.

3.1 Module BACnet: PID Controller

To check the validity of the control module, the current and instantaneous power estimates are compared with the installation data (see Fig. 4) on an operating day dedicated to testing the efficiency of the HVAC equipment. During the first week of testing, personnel have access to the air-conditioning system control panel for

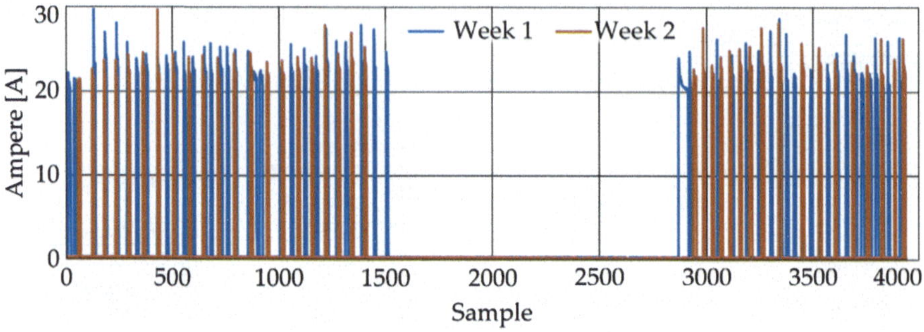

Fig. 6 Current consumption in amperes recorded in different test weeks. Comparison of the recorded current peaks. Blue line, normal compressor operation; red line, operation with BACnet module

Table 1 Comparisons of recorded measurements

Manual control	kWh current	kWh previous	kWh consumption	Cost kWh	Total dollars
	33.77	3.51	30.26	0.14	4.24
BACnet control and monitoring module	kWh current	kWh previous	kWh consumption	Cost kWh	Total dollars
	125.24	108.4	16.84	0.14	2.36

their day-to-day work. The air-conditioning equipment is switched on and off from 07h30 to 20h00.

In the blue signal Week 1, the amperes that the compressor of the air-conditioning equipment reaches are shown only with the control that the equipment has incorporated and also being manipulated by the users through the remote control. These starts reach an amperage of 25 amps. In the red signal Week 2, the control module is incorporated by BACnet; thus, users do not have access to the remote control, but through the previous programming, the control module assigns the comfort temperature according to the measurements and history generated. As can be seen in Fig. 6, in Week 2 the start-ups decrease, causing the consumption to decrease as well.

Table 1 and Fig. 7 show representative results of the evolution of the kWh consumption of the compressor incorporating the BACnet module. In this experiment, the day presents strong disturbances in solar irradiance and ambient temperature ranging between 30.6 and 33.0 °C. The set ambient temperature is 24 °C. Around 14h00 (sample 100) when user attention increases, the temperature starts to rise and the control acts by turning on the compressor. Once it stabilizes, the compressor is turned off, but due to the entrance and exit of users, the cold air is lost when opening the door around 15h30 (sample 159) and the compressor begins to act in several intercycles until 19h20 (sample 321), and then the attention is suspended until the next day. Figure 8 also shows the evolution of the temperature during the test using the BACnet module.

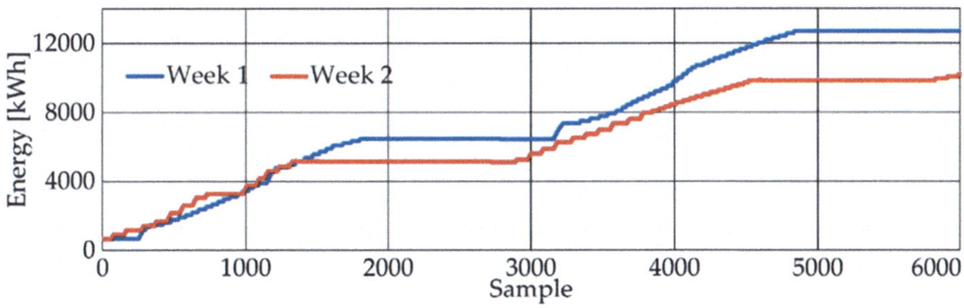

Fig. 7 Compressor kWh consumption on different test days. Blue line, normal compressor operation shows higher consumption than red line which is the operation with BACnet module

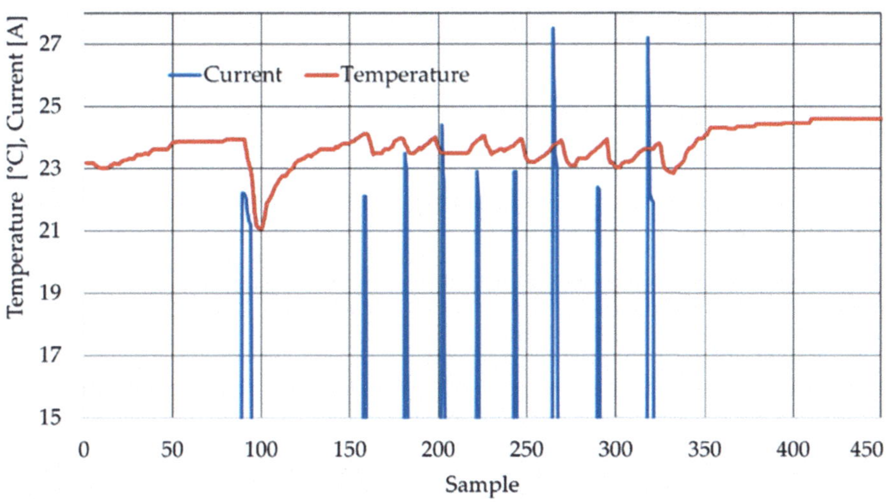

Fig. 8 Temperature in °C recorded by incorporating the BACnet module and start-up peaks measured at compressor start-up to maintain set temperature

Figure 5 shows the energy consumed on the test days. It can be seen that with module control (test Week 2), as the start-up times are more homogeneous and less frequent than with the manual control, the consumption is reduced.

4 Conclusions

This work shows the feasibility of using external control methods to those that usually bring air-conditioning systems commonly. The module is useful to analyze the different trends that occur in a space such as customer service where the dynamics of the attention there are times that the excess of people increases the temperature of the room or that the opening of the door loses cold air. From the tests, it was possible to obtain statistical data showing savings of up to 44% in 3 days out of 7 days of testing. Based on the tests performed, it can be concluded that the air-conditioning equipment are those that represent the highest power consumption in a building. One of the ways to achieve energy optimization within buildings is through the control and monitoring of these teams. It should be emphasized that this type of system has much more benefit when applied to machines that lack electronic technology in its components of higher consumption as motors and compressors. The network of air-conditioning systems currently existing through the BACnet protocol can be integrated with other types of systems, which seek to optimize energy; an example could be a lighting system.

Future work will focus on the use of BACnet modules developed for control design purposes and on the application of a hierarchical control strategy for the reduction of plant temperature.

References

1. Ortiz, J.G.: Diagnóstico y Optimización del consumo de Energía Eléctrica en el sector residencial, May, pp. 1–4 (2018)
2. Jiang, A., Zhu, Y., Elsafty, A., Tumeo, M.: Effects of global climate change on building energy consumption and its implications in Florida effects of global climate change on building energy consumption and its implications in Florida. Int. J. Constr. Educ. Res., 1–24 (2017). https://doi.org/10.1080/15578771.2017.1280104
3. Kameni, M., Reiter, S.: Evaluation of bioclimatic potential , energy consumption , CO2 - emission , and life cycle cost of a residential building located in Sub-Saharan Africa; a case study of eight countries. Sol. Energy. **218**(November 2020), 512–524 (2021). https://doi.org/10.1016/j.solener.2021.02.052
4. Coelho, G.B.A., Silva, H.E., Henriques, F.M.A.: Energy & buildings impact of climate change in cultural heritage: from energy consumption to artefacts' conservation and building rehabilitation. Energ. Buildings. **224**, 110250 (2020). https://doi.org/10.1016/j.enbuild.2020.110250
5. Rosiek, S.: Exergy analysis of a solar-assisted air-conditioning system: case study in southern Spain. Appl. Therm. Eng. **148**(October 2018), 806–816 (2019). https://doi.org/10.1016/j.applthermaleng.2018.10.132
6. Álvarez, J.D.: Repetitive control to improve users' thermal comfort and energy efficiency in buildings. Energies. **11**, 1–16 (2018). https://doi.org/10.3390/en11040976
7. Demarest, J.: HVAC-studie en automatisering kantoorgebouw (2019)
8. Royapoor, M., Antony, A., Roskilly, T.: A review of building climate and plant controls, and a survey of industry perspectives. Energ. Buildings. **158**, 453–465 (2018). https://doi.org/10.1016/j.enbuild.2017.10.022
9. Png, E., Srinivasan, S., Bekiroglu, K., Chaoyang, J., Su, R.: An internet of things upgrade for smart and scalable heating, ventilation and air-conditioning control in commercial buildings. Appl. Energy. **239**(January), 408–424 (2019). https://doi.org/10.1016/j.apenergy.2019.01.229
10. Cash, M., Wang, S., Pearson, B., Zhou, Q., Fu, X.: On Automating BACnet Device Discovery and Property Identification (2021)
11. Nast, M., Butzin, B., Golatowski, F., Timmermann, D.: Performance Analysis of a Secured BACnet/IP Network (2019)
12. Azman, M., Panicker, J.G., Kashyap, R.: Wireless daisy chain and tree topology networks for smart cities. IEEE Int. Conf. Electr. Comput. Commun. Technol., 1–6 (2019)
13. Edfors, O.: Decentralized massive MIMO processing exploring daisy-chain architecture and recursive algorithms. c, 1–14 (2020). https://doi.org/10.1109/TSP.2020.2964496

14. Morita, Y., Yamawaki, A.: Design of daisy-chain connection for flexible and scalable ultrasonic sensor array. 3, 3–7 (2018). https://doi.org/10.12792/icisip2018.027
15. Seem, J.E., Ph, D.: Implementation of a New Pattern Recognition Adaptive Controller Developed Through Optimization, vol. 103 (1997)
16. Seem, J.E.: A new pattern recognition adaptive controller with application to HVAC systems*. 34(8) (1998)

State of Decision-Making in the Baltic States: Nuclear Energy Past and for the Future

Guna Jakobsone-Snepste, Guillaume Deffuant, Ieva Vitolina, Janis Kapenieks Sr., and Valdis Vitolins

1 Introduction

On February 2, 2022, the European Commission (EC) approved in principle a Complementary Climate Delegated Act including, under strict conditions, specific nuclear and gas energy activities in the list of economic activities covered by the EU taxonomy [1] (the draft was formally adopted on March 9, 2022, and it was applied from January 1, 2023). With this document, the EC determines that the criteria for nuclear energy operations meet the EU's climate and environmental goals and will help accelerate the transition from solid or liquid fossil fuels to a climate-neutral future.

Currently, the Baltic countries do not use nuclear reactors for electricity production. Of the three Baltic states, only Lithuania has historical experience related to nuclear power plant (at Ignalina). However, both Estonia and Latvia have a future vision for a new nuclear power plant. How does the past influence today's scenarios. What is the overall direction for the Baltic states and what are the implications for nuclear energy production? It is not clearly approved at the governmental level. The reason is the difference of opinion between political parties and citizens. It is important to bring this topic up to date and find out what are the historical and current reasons for the fear of nuclear energy in the Baltic states. And why do the

The original version of this chapter was previously published without open access. A correction to this chapter is available at https://doi.org/10.1007/978-3-031-67987-2_26

G. Jakobsone-Snepste (✉) · I. Vitolina · J. Kapenieks Sr.
Riga Technical University, Riga, Latvia

G. Deffuant
INRAE, Paris, France

V. Vitolins
University of Latvia, Riga, Latvia

L. Chen (ed.), *Trends in Clean Energy Research*, Green Energy and Technology,
https://doi.org/10.1007/978-3-031-67987-2_14

Baltic countries need nuclear energy? If the Baltic states want to produce nuclear energy in the coming years, now is the last moment to start discussing this issue on a wider scale. It is also important to explain in more detail why the population is afraid of nuclear energy. Electric independence and very high electricity prices are the main reasons why this topic is very relevant for the Baltic states.

This paper is organized in three parts. The first part is the introduction, which describes Lithuania's nuclear power plant in Ignalina (1974–2009) and decommissioning (until 2038). The second part describes Visaginas' NPP project fate (2007–2012). The third part describes the current situation and plans, i.e., interconnection with European network and SMR, and finally, the conclusion.

2 The Ignalina Nuclear Power Plant (1974–2009) and Decommissioning (Until 2038)

Lithuania was not ready to close the Ignalina nuclear power plant, and in 1995, Jurgis Vilemas described the nuclear reactor's capacity. With the reestablishment of independence, Lithuania became another country possessing commercial nuclear reactors. In 1993, the Ignalina nuclear power plant (NPP) produced 12.26 billion kilowatt hours of electricity, which is equivalent to 87.1% of Lithuania's total electricity production. This is the highest annual nuclear share ever reached by any country. Economic conditions and the lack of viable alternatives demand Ignalina's continued operation. It is therefore vitally important to develop safety enhancements to reduce the associated risks. Lithuania has made good progress in generating a new nuclear industry infrastructure, and further strengthening it is essential [2].

The construction began in 1974, the first block of the Ignalina nuclear plant was completed in 1979. On December 31, 1983, the first reactor was started. Initially, it was planned to build three to four blocks. After the Chernobyl accident, the construction of the second block was postponed until 1987. In 1989, the third block was built. But after the "Ring of Life" protest, it was decided that the third block could not be built [3].

In 1987, I. Multer describes the risks of the Ignalina power plant as follows: The second RBMK 1500 unit at Ignalina in the Soviet Union recently become online. It has the same core dimensions as the Chernobyl RBMK 1000 units but a 50% higher power output. The higher fuel rating means that there would be less time for the operators to react in a loss of coolant accident. Also the reactor has design characteristics that would not be acceptable in the West [4]. Concerns about risks arose because in 1993, Lithuania set a world record for the share of nuclear-generated electricity produced in one country [5]. When Lithuania assumed control of the INPP (after the demise of the Soviet Union in 1991), the plant and its design and operational data have been completely opened and accessible to western experts. Initially, effective assistance in the nuclear safety field was provided by Sweden; subsequently, most states with significant nuclear expertise also contributed [5]. A significant conclusion stated in the SAR (Safety Analysis Report, 1995–1996) is that none of the analyzed safety concerns require the immediate shutdown of the plant [5]. It should be noted that SAR authors were specialists from Ignalina NPP, Russia, Canada, and Sweden [5].

The New York Times (1988) [5] wrote about a fire in a Lithuanian nuclear reactor. The newspaper reported that flames have been extinguished, the radiation has not escaped, and the reactor has been shut down after the accident. After this happened, fire safety measures were taken, and scientific articles were written about fire safety in Ignalina and fire hazard analysis. Stockholm's Swedish Defense Research Establishment said then it had registered four separate discharges of radiation from the power plant in 3 months [6] after Ignalina become online in 1983.

In 1997, an overview of safety issues of the Ignalina NPP was created, which briefly described the differences between Ignalina and Chernobyl. The consequences of the Chernobyl event were unique and were caused by a design flaw in the original RBMK-type plants. Subsequently, hardware changes were implemented to change the neutronic characteristics of the INPP (introduction of absorber rods, alteration of the fuel enrichment, and changes in the design of control rods). Therefore, INPP reactors cannot be equated with the unit, which failed at Chernobyl. The total power coefficient of reactivity should stay negative under all possible circumstances. This characteristic has been verified by extensive analysis by international experts [5]. Many (though not all) western reactors have been accomplished by a prominently visible, hemispherical shell. In the INPP, these functions were accomplished by an extensive system of interconnected re-enforced compartments called the accident localization system (ALS). This system uses the same principle as employed by the BWRs (boiling water reactors) built by GE (General Electric). Analysis has shown that a break in reactor pipes will not lead to the overheating of fuel and thus to the release of excess radioactivity [5]. The objective reasons for this are the higher complexity of an RBMK (comparable western BWR1s)-type reactor (a considerably larger number of pipes, valves, and associated equipment) and the lower level of quality control for Soviet design and construction. On the plus side, international analysis agrees that the INPP was remarkably robust and that vast majority of initiating events will not lead to fuel overheating and the release of radioactivity from fuel [5] (would not harm the environment).

Originally, the Ignalina plant was designed to provide power not only for Lithuania but also for neighboring countries such as Latvia and Belarus and the Russian exclave of Kaliningrad. In 1989, 42% of the power was exported, but this fell through the 1990s. In 1994, Lithuania agreed to accept funds—eventually 34.8 million ECU ($36.8 million) from the Nuclear Safety Account administered by the European Bank for Reconstruction and Development (EBRD)—to support a safety improvement program at Ignalina [7].

However, in 2009, the European Union said the plant was too dangerous to operate and that Lithuania has begun shutting down its main Soviet-era nuclear power plant [8]. Ignalina was the main producer of electricity in the country, and therefore there were many and radically opposite opinions expressed in the public space on this issue, for example, the opposite opinions of Lithuanian politicians. The EU said the facility was too dangerous. The reactors were based on the same models that failed at Chernobyl in 1986, producing the world's worst-ever nuclear accident [8]. The Lithuanian Prime Minister Andrius Kubilius states: "Lithuanians as decent Europeans, we are ready to meet our obligations to close the Ignalina power plant" [8]. Leonas Asmatas—former energy minister of Lithuania—see Ignalina's closure as a betrayal of Lithuania's security [8]. Ignalina's director, Viktor Shevaldin, considered it an unprecedented case when the country is left without a main electrical resource. There were even opinions that with the closure of Ignalina, Lithuania will become more dependent on Russian gas. This is especially controversial for some Lithuanians. Not only was their country occupied by Soviet Union during the Cold War, but Russia also has repeatedly shut off its gas pipelines in the price disputes with Ukraine in recent years [8].

The European Union was urging candidate countries to shut down their unsafe nuclear plants or take risk losing their bids to enter the EU [9]. Ignalina was the only plant with Chernobyl-style reactors still operating outside Russia. It closed its first reactor at the end of 2004 [8]. A new nuclear plant is scheduled to be built by 2020 [8]. Brussels offered EUR 45 million for closing Ignalina. Lithuania says it needs an additional 3 billion euros to cover all the costs involved in decommissioning the plant. Austrevicius says the calculation is based on the combined opinion of both international and local experts. The negotiator also urged caution, saying there is no previous experience in decommissioning Chornobyl-style reactors such as those at Ignalina [9]. In a study published by the Austrian Ecological Institute, it rated Ignalina plant as the third most dangerous plant, with a risk factor of 11. The two other plants deemed most unsafe—Armenia's "Armenia" plant and Bulgaria's Kozloduy plant—each received 13 risk points [9]. Presidential (Valdas Adamkus) spokeswoman Viola Gaizauskaite said: "The president started very clearly that what Lithuania has promised to do, it will. However, Lithuania cannot assume commitments that overstretch its financial and economic capabilities. Alone, Lithuania is not and will not be able to bear the cost of decommissioning" [9]. "Emma Udvin, a spokeswoman for the EU Commission in Brusssels, says the EU has already given Lithuania 100 million euros and is likely to give more—but not as much as Lithuania has requested" [9].

Until 2014, the European Union allocated 1450 million euros for the closure of the Ignalina nuclear power plant, 870 million euros were allocated for 2014–2020, and 610 million euros are planned for 2021–2029 [3]. After closing the Ignalina nuclear power plant, the decrease in the amount of electricity produced was −65.1%. The missing part was imported from Belarus and Latvia. After the closure of the NPP, Lithuania had to import 62% of the required amount of electricity. It was a severe test for the Lithuanian economy. Natural gas-based (fossil methane from Russia) CHP Lithuanian power plant became main plant, replacing electricity production at the Ignalina NPP after its closure in 2009. It shall be put into operation till 2012 for energy savings of 740 GWh by 2016 [10]. The first power unit of the power plant was stopped on December 31, 2004, and the second on December 31, 2009 [3]. The closure of Ignalina NPP marked the turning point for the Lithuanian energy sector [10]. Although the power unit of the power plant was stopped, decommissioning works are still ongoing. After the construction of a new spent nuclear fuel storage facility in 2017, it stores the spent fuel of the power plant. After 2020, the dismantling of the power plant buildings began. Decommissioning works are planned to be completed in 2038 [3].

The main policy document to promote energy efficiency in Lithuania is the National Energy Efficiency Programme for 2006–2010. The Energy Efficiency Action Plan for 2010–2016 will also be implemented [10]. Currently, the National Energy and Climate Action Plan of the Republic of Lithuania for 2021–2023 foresees to safely decommission the Ignalina nuclear power plant and dispose of the resulting radioactive waste [11]. Unfortunately, in 2021, Lithuania had remained systematically and infrastructurally linked to the Russian energy system. Considering that it did not have energy interconnections with the EU mainland and that all fossil methane and most of electricity were purchased from a single monopolistic supplier, Lithuania started implementing infrastructure projects of regional importance to interconnect the Lithuanian and EU energy systems [11]. The National Energy and Climate Action Plan of the Republic of Lithuania for 2021–2023 provides reducing the influence of a single (monopolistic) energy supplier: by increasing the use of local and renewable resources, developing competitive local energy generation capacities, and diversifying imports of energy resources that cannot be replaced by local sources [11]. In 2021, the situation of the Lithuanian electricity market was as follows: the electricity system is connected to the Belarussian, Russian, Latvian, Polish, and Swedish electricity systems [11]. The Lithuanian-Latvian-Estonian-Belarussian electricity systems are part of the IPS/UPS system operated by the Russian electricity system operator [11].

In order for Lithuania and the Baltic states to move toward electricity independence, much attention is paid to the synchronization project (it is defined in all strategic documents), which would synchronize the electricity systems of the Baltic states with the continental European network. Work on the extension of the LitPol Link interconnection is also ongoing. A high-voltage 500 MW DC southwest interconnection—LitPol Link—costing EUR 250–300 million, to improve the transmission capacity between Lithuania and Poland, was commissioned in 2015 [7]. This follows the inauguration of an interconnector between Estonia and Finland to the north—Estlink-1, a 150 kV, 350 MW DC cable costing EUR 110 million and also supported by EU funding [7]. Another major transmission link westward under the Baltic Sea, the 700 MWe NordBalt project, connects Klaipeda in Lithuania and Nybro in Sweden. A 550 million project was completed in 2016 [7]. The revised energy policy in 2012 involved rebuilding the grid to be independent of the Russian/Belarus system and to work with the European Network of Transmission System Operators (ENTSO) synchronous system, as well as strengthening interconnection among the three Baltic states [7]. Lithuania continues to take measures to isolate Kaliningrad [7].

In 1958, the USSR Council of Ministers gave a permission to build a typical research nuclear reactor (IRT) in the vicinity of Riga [12]. The "basin-type" typical research nuclear reactor (IRT) with a 1000 kWth heating capacity was built in *Salaspils*—20 km from Riga, Latvia. The IRT was used mainly for scientific research in radiation chemistry, nuclear physics, and radiobiology. The most significant work performed on the critical assembly was the unique liquid metal regulator of the nuclear reactor intended for the use in nuclear reactors in a weightless state in space. Rotary cylindrical regulators of nuclear reactors were also investigated [12]. In the early 1950s, the idea of creating an IRT arose, and on September 26, 1961, the Salaspils reactor was launched. Afterward its heating capacity was increased from 2000 kWth, after the reconstruction in 1974, to 5000 kWth [12]. The results obtained during the research were not applied directly in nuclear energetics. At the end of the 1950s, the Latvian scientists gained remarkable achievements in the application of radioactive isotopes and nuclear radiation [12]. The reactor continued working till its stoppage in 1998. On the whole, it had been active for 37 years and consumed 12.3 kg of the uranium 235 U isotope [12]. Latvian scientists, together with specialists from Moscow, developed a project for the expansion of block 3 of the Ignalina NPP with a radiation contour.

In general, the currently performed and planned *closing cost of the Ignalina nuclear power plant is EUR 2930 billion.* The total expenditure should also count the *additional security investment of USD 36.8 million* from the Nuclear Security Account managed by the European Bank for Reconstruction and Development to support the safety program at Ignalina.

3 Visaginas NPP Project Fate (2007–2012)

Visaginas nuclear power plant was a planned nuclear power plant project in Lithuania. It was proposed to be built at the site of the closed Ignalina nuclear power plant [13]. The plan was to build in 2020 a new nuclear power plant (Visaginas NPP) together with Estonia. The planned Visaginas nuclear power plant constitutes an integral part of the Baltic Energy Market Interconnection Plan (BEPIM) [7]. BEPIM was signed by eight states of the Baltic region and the European Commission [7]. *The Visaginas power plant has not moved forward following a referendum in October 2012 in which two-thirds voted against the project proceeding* [7]. The year 2010 was the first year without nuclear power [10] at Lithuania.

In 2009, the government commissioned an international advisory consortium, led by the Rothschild investment bank, to prepare a business model and financing plan for the Visaginas project. This was adopted by the government, which proceeded to invite expressions of interest by major strategic investors to supplement the role of regional partners in providing a plant up to 140 MWe. This was put forward as an attractive investment proposition, with favorable environment, infrastructure in place, support from the EU Commission, and competition from Kaliningrad as the only downside. The investor would get a majority stake (probably 51%) in the proposed new plant, alongside Lithuania's Lietuvos Energija, Latvia's Latvenergo, Eesti Energia, and (until December 2011) Poland's Polska Grupa Energetyczna (PGE)i. However, with Lithuania wanting 34% of the project and Poland wanting 30% of it, Latvia and Estonia were unhappy with the prospect of minor stakes and the split was far from clear [7].

GE Hitachi planned to build a single 1350 MWe ABWR, several of which are operating in Japan, or a pair of these. The cost of the project was estimated at *EUR 4.92 billion* [7].

Although at an early December 2010 meeting in Warsaw, prime ministers of Lithuania, Latvia, Estonia, and Poland confirmed their support for the Visaginas project [7], the further and final decision on the non-implementation of the Visagina project was also influenced by the annexation of Crimea (Ukraine).

There was a 44.6% inflation (all goods and services) increase in February 2024 compared to May 2015 in Latvia [14]. In Lithuania between 2015 and today, inflation produces a cumulative price *increase of 55.16%* [15]. If we assume that the construction costs of Visaginas will be more than twice (coef. 2.2) as high in 2027 and increased labor costs, then the construction costs could be EUR 10.82 billion for 3400 MWe project: (1 + 0.5 (labor costs) +0.7 (materials)) =2.2 (coefficient).

4 Current Situation and Plans: Interconnection with European Network and SMR

The fact that energy independence is important for *Latvia is confirmed by* the established Ministry of Climate and Energy, which was created on January 1, 2023. And in January 2024, the informative report "Nuclear Energy Development Opportunities in Latvia" has been currently under discussion and not approved by the Cabinet of Ministers. In order to achieve climate goals and increase electricity production capacity, the development of the nuclear power industry is consider as one of the solutions, planning the construction and long-term use of nuclear power plants with small modular reactors (SMRs). Considering that all technical solutions are in the initial stage of development, their implementation costs as well as benefits are not yet clearly known [2].

If a decision is made in Latvia on the use of technical solution to develop the nuclear energy sector, it can be implemented in two different ways: (1) build one or more nuclear power plants and (2) look for cooperation opportunities to develop the nuclear energy industry in cooperation with Estonia [2]. In a situation where both Latvia and Estonia evaluate the development of nuclear energy as potentially the best solution for ensuring energy supply in the future, cooperation between the two countries is theoretically possible [2]. The construction of a nuclear power plant in cooperation with Estonia would solve the issue of storage of used nuclear fuel—in the east of Estonia, the structure of the earth's layers is suitable for long-term storage of used nuclear fuel, while this is not possible in the rocks of Latvia [2]. Taking into account that the preliminary investments of the nuclear power plant reach 4,500,000 euro/MW, then approximately 2 billion euros is needed for the construction and feasibility study of the SMR nuclear power plant with a capacity of 300 MW [2]. A cooperation project with Estonia would be more beneficial for Latvia (in the 2022, the electricity produced in Latvia had a deficit of 2.311.527 MWh) [2].

Table 1 LCOE assessment of potential power generation resources [2]

Electricity production technology	Capital costs (*euro*/kW)	Operating costs (*euro*/kW)	Power (MW)	Workload (%)	Life cycle (years)	LCOE estimate (*euro*/MWh)
Solar	925	20	100	18	25	60–80
Hydroelectric power station	2902	50	100	50	80	60–80
Onshore wind farms	1931	40	300	35	25	65–85
Nuclear power SMR	4500	128	300	80	60	*70–95*
Offshore wind farms	2740	86	300	40	25	90–110
Thermal power plant with a combined cycle gas turbine	1092	475	300	65	30	100–120

Latvia's electricity policy has undergone drastic changes. Already in 2022, the electricity trade with Russia has been terminated, and from July 2023, electricity trade transactions with the Russian Federation and the Republic of Belarus, as well as the supply of electricity for trade using the electricity system of the Russian Federation and the Republic of Belarus, are prohibited [2].

To assess the economic competitiveness of energy, including electricity generation technologies, the levelized cost of energy (LOCE) is used as a generally accepted comparison indicator, which is determined taking into account the investments and operating costs required for the construction of the power plant, as well as the life cycle of the plant and the electricity produced during this cycle [2]. The report's LOCE calculation indicates that modular (SMR) nuclear power plants are generally moderately expensive compared to other power generation technologies (Table 1).

The fourth-generation nuclear reactors retain the ability of nuclear reactor to quickly adjust electricity production to the actual demand, but since the SMR capacity is significantly lower than the production capacity of traditional nuclear reactors, the reactor is operated with a much smaller amount of nuclear fuel (enriched uranium) [2]. The time required from the moment when a decision has been made on the production of electricity in nuclear power plants to the start of operation of the first nuclear power plant is approximately 10–15 years [2]. In 2010, as part of the conceptual study construction of nuclear power plants in Latvia, "Latvenergo" examined six potential locations for new nuclear power plants—three possible locations on the coast of the Baltic Sea (West of Latvia), namely, Pape, Pāvilosta, and Ovīši, and two locations near a large lake, i.e., Burtnieki and Rāznas, and one location near Riga—Acone [2] (Fig. 1). In order to make a decision on the construction of a new nuclear power plant with SMR in Latvia, the location of the plant should be revaluated [2]. The conditions for the construction of a nuclear

Fig. 1 Preliminary location of nuclear power plants according to "Latvenergo" 2010 research data [2]

power plant in *Estonia* will be evaluated in accordance with the final report of the nuclear energy working group on the prospects for the development of nuclear energy, which was published on the website of the Ministry of Climate of Estonia in the Estonian language "Tuumaenergia töörühma lõpparuanne" on December 30, 2023 [2]. High-level representatives of state institutions participated in the working group, and the report was developed in accordance with the guideline developed by the IAEA "Cornerstones in the development of the national infrastructure of nuclear energy" [2]. The main international cooperation partner in all topics related to radiation and nuclear safety in Estonia is the IAEA, which offers various trainings, consultations, and expert missions and compiles guideline materials and international standards. The USA, France, Germany, Japan, Canada, and Finland have also offered cooperation and assistance in the field of nuclear energy [2]. For now, there is hope that Estonians will vote for SMR.

The most suitable technology for a possible Estonian nuclear power plant, according to the research conducted by Fermi Energia OU, is the boiling water MWRX-300 nuclear reactor developed by GE Hitachi. It is estimated that the total capacity of the nuclear power plant could reach 600 or even 1200 MW (depending on the selected development scenario). According to surveys conducted by Fermi Energia OU, at the beginning of 2022, the construction of a nuclear power plant in Estonia was supported by 59% of the surveyed population [2].

Lithuania has developed plans for its energy independence and greater energy efficiency and has been actively working on the possibility of developing the use of its natural resources in order to become less dependent on imports. And at the same time, the energy policy is created in such a way that in the future greenhouse gas emissions are reduced and the use of renewable energy sources is increased.

Lithuania is currently carrying out research to assess the future use of nuclear energy and greenhouse gas emission monitoring methodologies, including energy production with SMR [2].

Estonia is highly dependent on fossil fuels, especially oil and electricity produced from these sources. However, Estonia is developing rapidly in the field of alternative energy sources such as wind and biomass. Also, Estonia is actively working on the possibility of developing the use of its natural resources in order to become less dependent on imports. And at the same time, the energy policy is created in such a way that in the future greenhouse gas emissions are reduced and the use of renewable energy sources is increased, including the use of SMR.

Like other Baltic states, *Latvia* was also highly dependent on imported energy resources, especially Russian gas. However, Latvia is producing large part of green electricity with hydroelectric power plants and is strongly focused on biomass, wind, and solar energy as sustainable sources of energy. Latvia has taken measures in the energy sector to promote more efficient energy supply and increase energy supply security. And at the same time, the energy policy is created in such a way that in the future greenhouse gas emissions are reduced and the use of renewable energy sources is increased. Latvia has set high goals regarding the use of renewable energy sources and the reduction of emissions in order to achieve the goals set by the European Union and has started discussions in the political environment about the possibilities of SMR. The nuclear power plant construction projects planned in Latvia were stopped along with the Lithuanian protest action "Ring of Life." Society is not yet ready to accept the construction of new NPP in Latvia; therefore, it is important to update this topic and introduce the differences and technological progress that distinguish Ignalina-type reactors from the new modular reactors (it is necessary to educate about the benefits and the extent of the risks). In the summer of this year, the first results of Latvian-Estonian cooperation opportunities in the field of wind energy production will be known.

5 Conclusions

The measurement, assessment, and effective mitigation of energy intensity compose a foremost objective of contemporary energy policy. This is because energy is the main intermediate resource for the socioeconomic development in any country [2]. Consequently, the appropriate energy use enables to resolve the problems of economic competitiveness, energy security, and environmental sustainability [2]. The Baltic Energy Market Interconnection Plan (BEMIP), endorsed on June 17, 2009, is another essential factor driving energy infrastructure projects further. The BEMIP, an initiative of the European Commission with eight participating Baltic Sea states, is an unprecedented step in the EU energy policy [10]. All three countries are working to diversify their energy sources to reduce dependence on foreign resources, especially Russia. Considering that the development of SMR technologies is very fast, in the next 10–20 years, SMR technologies can become a widely applicable and effective solution in the production of electricity and thermal energy [2]. SMRs are fourth-generation nuclear reactors with increased safety requirements, higher efficiency, and reduced amount of radioactive waste [2]. The advantage of SMRs is their simplified design, economy in case of mass production, and possibility of completely building the modules in the factory, delivering them to the developer by rail or using a truck. The principle of passive cooling also plays an important role, which significantly increases the safety of SMR [2]. Therefore, it is necessary to continue the research in this field and to educate and inform consumers about the benefits and risks associated with SMR. Fears of a repeat of the Chernobyl disaster must be allayed. It should be noted that until now the Chernobyl disaster has not been repeated in reactors similar to it. The Baltic states must continue to develop innovative technologies and increase their energy independence.

Acknowledgement This work has been supported by the European Union HORIZON programme, Project number: 101079206, Project acronym: TED4LAT, Call: HORIZON-WIDERA-2021-ACCESS-03.

References

1. European Commission: EU taxonomy: Complementary Climate Delegate Act to accelerate decarbonization. In: European Commission, Business, Economy, Euro, Finance, European Commission (2022)
2. Latvia's Ministry of Climate and Energy: Nuclear Energy Development Opportunities in Latvia. In: Ministry of Climate and Energy, pp. 2–20. Ministry of Climate and Energy, Riga (2024)
3. General Lithuanian encyclopedia Homepage, http://www.vle.lt. Last accessed 2024/03/17
4. Multer, I.: Exploring the safety of Ignalina. Nucl. Eng. Int. **18**(23), 21–22 (1987)
5. Almenas, K., Uspuras, E.: A Brief Overview of Ignalina NPP Safety Issues. INPP Printing Department, Lithuania (1997)
6. Iams, J.: Soviets use advanced system after fire at nuke plant. In: UPI, News World Communications, Washington, D.C. (1988)
7. World Nuclear Association Homepage: http://world-nuclear.org. Last accessed 2024/03/17
8. Houlton, S.: Shutting down. In: Deutsche Welle, Reuters, Bonn (2009)
9. Mite, V.: Lithuania: EU urges closure of Ignalina nuclear plant. In: Radio Free Europe. Radio Free Europe, Washington, D.C. (2002)
10. Baležentis, A., Baležentis, T., Streimikiene, D.: The energy intensity in Lithuania during 1995–2009: a LMDI approach. Energy Policy. **39**(11), 7322–7334 (2011)
11. Lithuania Government: National energy and Climate action plan of the republic of Lithuania for 2021–2030. In: European Commission, Energy. Climate change, Environment, pp. 8–39, European Commission (2022)
12. Mikelsons, K., Ekmanis, J., Gavars, V., Tomsone E., Zeltins N.: Development of nuclear energetics in Latvia. In: U.S. Department of Energy Office of Scientific and Technical Information (2010)
13. Pajuste, E., Līckrastiņa, A., Bajinskis, A., Stīne Teimane, A.: Nuclear education for sustainable nuclear energy development—past and future perspective of Latvia. In: 20th INPRO Dialogue Forum on Challenges and Issues in Capacity Building for Ensuring Nuclear Energy Sustainable, Development. OAK Ridge National Laboratory (2023)
14. Inflation Calculator of Latvia Homepage. http://tools.csb.gov.lv. Last accessed 2024/03/17
15. Inflation Calculator of Lithuania Homepage. http://officialdata.org. Last accessed 2024/03/17

The Transition to 100% Renewable Energy Versus the Global Temperature Scenarios: A Perspective Analysis

Joseph Akpan and Oludolapo Olanrewaju

1 Introduction

The world is projected to experience higher levels of global warming compared to the current global average. It is primarily due to socio-economic factors, demographic trends, dependence on fossil fuels, and a slow adoption of renewable energy measures [1, 2]. Consequently, many nations are particularly more susceptible to the impacts of climate change [3, 4]. However, the effect of delay in the high penetration of renewable energy transcends beyond just the affected countries as the impacts of climate change are anticipated to influence the rate at which glaciers melt, alter weather patterns, and modify the intensity and timing of the monsoon [4–6]. These changes have significant implications for fossil fuel-based energy systems.

In recent years, renewable energy has become the favoured approach for power generation, surpassing fossil-based methods, owing to the significant cost reductions in renewable energy technologies [7]. The IPCC reports recommend that renewable energy sources in the global energy generation should make up 52–67% of the world's primary energy supply by the year 2050 [8]. According to another study on the role of the International Renewable Energy Agency (IRENA) in the energy transition to 100% RE, renewable energy sources are projected to make up nearly 66% of the world's total energy consumption during the transition towards limiting global warming to below 2 °C [9].

Adopting a predominantly renewable energy system offers numerous benefits, most particularly in maintaining the right global temperature averages and other advantages such as addressing the problem of fossil fuel exhaustion, ensuring energy supply reliability, stabilising energy costs, and reducing pollution and health risks. Therefore, several efforts have been made by both the academic and the industry in

J. Akpan (✉) · O. Olanrewaju
Industrial Engineering, Durban University of Technology, Durban, South Africa
e-mail: josephsamuelakpan@gmail.com; oludolapoo@dut.ac.za

L. Chen (ed.), *Trends in Clean Energy Research*, Green Energy and Technology,
https://doi.org/10.1007/978-3-031-67987-2_15

this regard. These efforts have made the history of the development of the 100% RE idea, which was discussed extensively in the work by Akpan et al. [7]. The progress and growth in 100% RE also seem to provide guiding assurances to develop policies that drive this endeavour. Topical research is currently prioritising the expansion of renewable energy capabilities and investigating the feasibility of attaining a 100% reliance on renewable energy sources. In Sect. 2, the review of key studies in 100% renewable energy are reviewed; alongside this study, research problem is identified and discussed with the rationale. This study is structured into five sections. Section 1 presents the overview and history of the 100% renewable energy system, and Sect. 2 states the problem statement and the rationale behind the study. In Sect. 3, the methodology and hypothesis development are explained, while the results and discussions are included in Sect. 4. Finally, the limitation and conclusion of the work are made in Sect. 5.

2 Problem Statement and Study Rationale

The definition of 100% renewable energy systems continues to vary across several literatures. However, the common understanding has been in continuous increment of the share of renewable energy in the global energy system to ensure there is the least effect across all sectors (power, building, transportation, industrial, and others). Consequently, a definition was made by Akpan J. and Olanrewaju O. in [10] that the transition to 100% renewable energy represents a substantial transition in the global energy sector, seeking to substitute all fossil fuels and other non-renewables with sustainable alternatives. The summary of the studies in Table 1 discusses various studies on 100% renewable energy (RE) systems in different countries and highlights the deficiencies in those studies, forming the rationale for this study.

There is a dearth of studies that define the clear path towards the 100% RE transition in line with the global temperature levels. The summary of literature presented in Table 2 is conducted in terms of this study's research emphasis on the existing gaps in many studies that often fail to link the 100% RE possibilities with global average temperature scenarios. Some studies examined the whole energy system (including transportation, heating/cooling, and industry), while others concentrated on the power sector (and, more especially, the shift to a renewable energy system that relies entirely on renewable sources). For obtaining a thorough evaluation of the present study, both the power sector and the whole energy system resulting from the global primary energy mix are considered. Within a strong climate policy framework, climate scenario targets should be aligned with exploiting renewable energy goals.

Table 1 Summary of key 100% renewable energy studies in top selected CO_2 emitters

Country	Summary of studies	Target year for 100% RE	Climate scenario
China	One of the earliest experimental projects into 100% RE, suggesting a 100% RE system analysis for China	N/D	No
China	Design optimisation is suggested for improving 100% renewable energy systems in low-density areas	N/D	No
China	This Beijing study used two-phase energy system models to study Beijing's 2030 energy market reaching 100% RE	2030	No
USA	100% renewable energy (RE) in US electric power networks were simulated. The least cost buildout reaches 57% RE penetration in 2050 under base conditions and USD 39/MWh at 100%	2050	No
USA	An optimisation problem is proposed to determine the amount of renewable generation and energy storage needed to balance 100% of a utility's hourly electricity demand	2040–2045	No
India	The model optimises the least cost combination of RE power plants and storage technologies to create a completely RE-based power system by 2050 based on 2015 installed power plant capacities, lives, and total energy demand, enabling a 100% renewable energy system	2050	No
India	Delhi's 100% renewable energy system's technological and economic potential is examined in this study	N/D	No
Europe, Eurasia, and MENA regions	This study explored the feasibility of a regional integrated renewable energy-based carbon-neutral power system using existing energy generation, storage, and transmission technologies throughout Europe, Eurasia, the Middle East, and North Africa with a total LCOE of about EUR 42/MWh	2030	No
Japan	The research designed and evaluated a renewable energy system for Akita, Japan, using wind power potential and batteries to produce hydrogen, cutting electricity costs by 57% and overall costs by 32%	N/D	No
Japan	Batteries met demand when other energy sources failed. The "no biomass", "supply shortage", and "baseload" situations were explored. Compared to "no biomass" electricity prices, "baseload" lowered them all		No

(continued)

Table 1 (continued)

Country	Summary of studies	Target year for 100% RE	Climate scenario
Japan	Japan's renewable energy future using a 40-year hourly energy balance model was examined	2050	No
Germany	The study tried to clarify the possibility of Germany's 100% renewable energy transition in 2050	2050	No
Germany	This research examines Germany's 100% renewable and sector-coupled energy system's viability	N/D	No
Iran	The report forecasts the possibility of 100% renewable in Iran by 2050	2050	No
Canada	This article evaluates the infrastructure costs for transitioning to carbon neutrality for Canada's ten provinces until 2060	2060	No
South Korea	The research develops a renewable energy forecasting model using the Korean energy policy as a case study. It analyses four renewable energy scenarios using deep-learning-based models to anticipate power demand and generation	Annual	No
Indonesia	This study investigates Indonesia's 2050 100% renewable energy power system transition using TIMES' least cost optimisation	2050	No
Saudi Arabia	The first research study indicates that by combining the electricity and growing desalination sectors, Saudi Arabia can achieve a 100% renewable energy power grid by 2050	2050	No
Saudi Arabia	As a follow-up to the first Saudi Arabia 100% RE study, the new study presents that a full transition to renewable energy can be possible if single-axis tracking PV and battery storage are the system's primary LCOE drivers	2040–2050	No
South Africa	South Africa's energy transition is simulated hourly until 2050. The findings imply that solar PV and wind energy can replace coal in electricity	2050	No
Egypt	Egypt's wind energy potential is understudied, so the author examined two 300 MW wind farms for roughness factor and wind power density	Annual	No
World	Contribution of this study, simulation 100% RE against global temperature scenarios	2100	Yes (This work)

Adapted and modified from [10]
N/D not defined

Table 2 Comparison of the scenarios. Source: author's elaboration

Scenario	CO_2 emissions	Renewables subsidy	Growth in renewable share
BP	High	None	Business-as-usual
OP	Low	Moderately subsidised	Approximately 50%
MOP	Lower	Highly subsidised	Approximately 55%
EOP	Negative emissions	Very highly subsidised	Approximately 100%

3 Methodology

3.1 Model Setup

Many different types of energy consumers and producers, including those in the industrial, commercial, residential, and agricultural sectors, have been accounted in the EN-ROADS model, as shown in Fig. 1. The data flow used in this study is shown in the schematic diagram of Fig. 2, which incorporates the four key sectors that heavily drive CO_2 emissions towards the complete integration and transition into 100% RE. Figure 1 depicts the architectural framework of the EN-ROADS energy transition.

The EN-ROADS model is utilised for the global transition research from 2023 to 2100 in 1-year time steps. The model primarily optimises energy system parameters under constraints and assumes future average global temperature rise and CO_2 emissions. However, this study is also interested in the energy mix that drives this output of global temperature rise and emissions. Hence, the energy mix result is also evaluated to allow the understanding of what amount of energy per technology and the value of renewables needed to achieve the desired climate scenarios. The main usefulness of EN-ROADS over other models is that it offers transition pathways that allow for the iteration of different policy scenarios and strategies to simulate and obtain results based on Eq. (1). The post-processing of simulation results is also performed to make the comparison:

$$\boldsymbol{T_E} = f\left(E_S,\ T_{\text{ee}},\ B_{\text{ee}}, I_{\text{ee}},\ G_{p,\text{GDP}}, \text{LFI}_e,\ C_r\right) \tag{1}$$

From Fig. 2, the transition into 100% RE is studied against the effect on global temperature levels T_E using variable input data from sectoral contributions (power, i.e. energy supply, transport, buildings, and industry), growth, emissions, and emission removal technologies, where E_S is the energy supply from all the energy sources (renewables, oil, coal, natural gas, nuclear, and bioenergy). T_{ee}, B_{ee}, and I_{ee} are the electrification and energy efficiency for transport, building, and industry sectors, respectively. $G_{p,\text{ GDP}}$ is the growth in population and GDP. LFI_e is the land, food, and industry emissions, while C_r is the CO_2 removal rate.

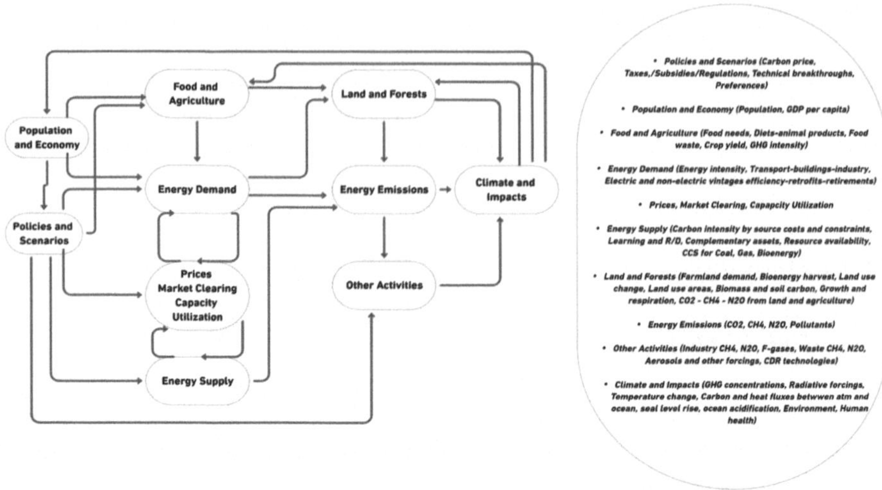

Fig. 1 The general model of energy-emissions-climate with EN-ROAD. Modified from [11]

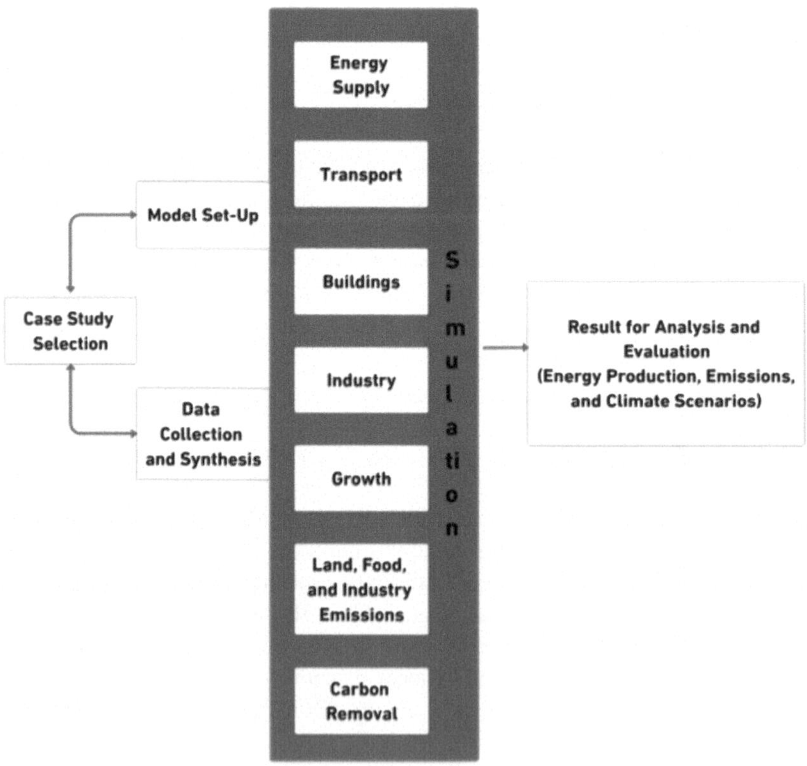

Fig. 2 The EN-ROADS model from input-output

The change across these sectors, coupled with growth in population and the GDP, determines the amount of CO_2 (CO_2 atm) that goes into the atmosphere. The relation is computed using the most used mathematical identity by IPCC in emission computation; the Kaya Index is expressed as in Eq. (2):

$$CO_2\ \text{atm} = P * \frac{\text{GDP}}{P} * \frac{E_i}{\text{GDP}} * \frac{CO_2\ \text{atm}}{E_i} \tag{2}$$

where P is the population, GDP is the expression of economic growth, and E_i is the intensity of energy from across the different energy sources based on production and anticipated consumption. The temperature values can then be computed using Eqs. (3) and (4):

$$CO_2 = \frac{\mathbf{CO_2\ atm}}{\mathbf{CO_2\ pre - industrial\ atm}} \tag{3}$$

$$T_E = \frac{\kappa}{\lambda} * \frac{\mathbf{In}\left(\frac{\mathbf{CO_2\ atm}}{\mathbf{CO_2\ pre{-}industrial\ atm}}\right)}{\mathbf{In\ (2)}} \tag{4}$$

where λ and κ are the climate feedback parameters and heat transfer constant, respectively.

3.2 *Assumptions*

3.2.1 Scenario Selection

In order to properly examine the potential transition pathways, four scenarios were introduced and considered. The following provides a detailed summary of these scenarios:

i. Baseline Path (BP) Scenario

This scenario assumes that the current energy policy and mix would continue and continuous heavy reliance on imported fossil fuels, particularly coal and natural gas, for electricity generation. The renewable sector would also play a role, but the production of energy from renewable would be slow. In the BP scenario, little or no tax on carbon causes slow improvement in energy efficiency and electrification at the present status quo. As a result, and without carbon removal technologies, CO_2 emissions from the energy sector are expected to continue to rise.

ii. Optimistic Path (OP) Scenario

This scenario assumes the implementation of policies (medium carbon tax, less energy efficiency improvement, less electrification, and subsidised utilisation of carbon removal technologies) that aim to increase the share of renewable energy in the global energy mix. By 2050, the policy aims to generate 50% of the global primary energy from renewable sources and with at least a share of 3% in energy storage. Hence, a significant increase in the deployment of solar, wind, and biomass power plants would be required. In order to achieve these targets, the policy also calls for the retirement of all coal, nuclear, natural gas, and oil-based power plants after 2030 and an increase in energy efficiency, at least by a factor of 2.5.

iii. More Optimistic Path (MOP) Scenario

This scenario assumes that the world would implement the IPCC recommendation of keeping the global temperature less than 2 °C, which aims to reduce energy production from fossils by 45%, increase renewable share by 55%, and with at least a share of 3% in energy storage. The energy efficiency is expected to increase by at least a factor of 2, and high electrification is achieved. Widespread adoption of energy-efficient technologies and practices across all sectors of the economy is essential. By reducing energy consumption through high-energy efficiency of at least 3, the burden on the electricity grid would be eased, and there would be less needed to build new power plants.

iv. Extreme Optimistic Path (EOP) Scenario

This scenario assumes that the globe would achieve 100% of the primary energy from the combination of renewable energy and energy storage, which aims to reduce the emissions of CO_2 to net zero. This scenario is highly ambitious and would require widespread adoption of energy-efficient technologies with an increase in energy efficiency by at least a factor of 4 and very high electrification across all sectors of the economy. By reducing energy production from fossil-based fuels, the burden on the atmosphere and climate would be eased. A summary of the different scenarios is compared and presented in Table 2.

3.2.2 Key Input Parameters with Explanation of Assumptions Used in the EN-ROADS Model

The simulation of the transition in this study based on the EN-ROADS model involves primarily four main categories of techno-economic and socio-economic parameters: tax/subsidy in future electricity generation technologies, technology breakthroughs, reduction in non-renewable energy infrastructures, carbon pricing, energy efficiency, electrification, population/economic growth, afforestation/deforestation, and carbon removal. These strategies are presented in ref. [12] with the corresponding indicators.

The techno-economic and socio-economic elements are taken into consideration to anticipate energy consumption, production, carbon dioxide emissions, and climate scenarios. The input parameters per module are described in ref. [12].

The energy supply module uses synthetic data as the baseline input to compute energy demand, considering past trends in the production, demand, and consumption of coal, oil, natural gas, bioenergy, nuclear power, and renewable energy sources

[13, 14] which are then used to obtain the scenarios for the OP, MOP, and EOP, respectively. The cost of various energy sources has a considerable influence on decisions about energy infrastructure [15]. Hence, the cost of coal directly affects investments in new capacity, while the cost of oil directly affects investments in new drilling operations and refineries. Additionally, bioenergy sources are subject to distinct subsidies or taxes. The renewables indicator monitors the duration of wind and solar installations as they go through various phases [16], whereas nuclear power plants monitor the phases of their development up to energy production and emerging environmental justice issues [17]. The carbon price extracted from the World Carbon Pricing Database [18] shows that carbon price exclusively impacts the emissions of carbon dioxide originating from energy sources. To modify emissions other than the baseline values [13, 14], managing the deforestation and afforestation indicators influences the changes. The cost of coal, oil, and building energy-efficient energy systems would directly impact emissions from transportation, construction, and industrial activities. However, enhancing the energy efficiency of buildings and industries would progressively reduce emissions by considering the collective average of all infrastructure [19, 20].

Emission removal technologies play a vital role in mitigating greenhouse gas emissions, while the net CO_2 removals are diminished compared to the total removals owing to the natural processes of decay and forest fires in older or diseased forests. The maximum capacity for technical carbon removal is determined by the median value of the ranges shown in the 2018 'Greenhouse gas removal' research conducted by the Royal Society [21]. The primary factor that influences the financial appeal of electric cars and related infrastructure is the integration of transport electrification. The model framework also monitors the total efficiency, encompassing the modification of existing assets.

Economic growth is a significant consideration in the EN-ROADS model, as it predicts that the GDP per capita growth in all regions would gradually align with a sustainable long-term economic growth rate of at least 1.5% per year. The model considers the influence of climate change on GDP, resulting in a long-term growth rate in the baseline scenario that is below 1.5% per year. The calculation of total global GDP, also known as gross world product, involves multiplying the population (data from the United Nations World Population Prospects in [22]) by the economic growth rate, which is measured by GDP per capita.

4 Result and Discussion

From 2023 to 2100, the results estimate the global primary energy mix by technology under the four scenarios, as well as the CO_2 emissions and the corresponding changes in the average global temperature level by 2100. The profile of the energy resource per technology is made based on four scenarios, where the projection of energy production based on the BP, OP, MOP, and EOP scenarios are such that the energy mix is continuously being dominated by renewables, less reduction in

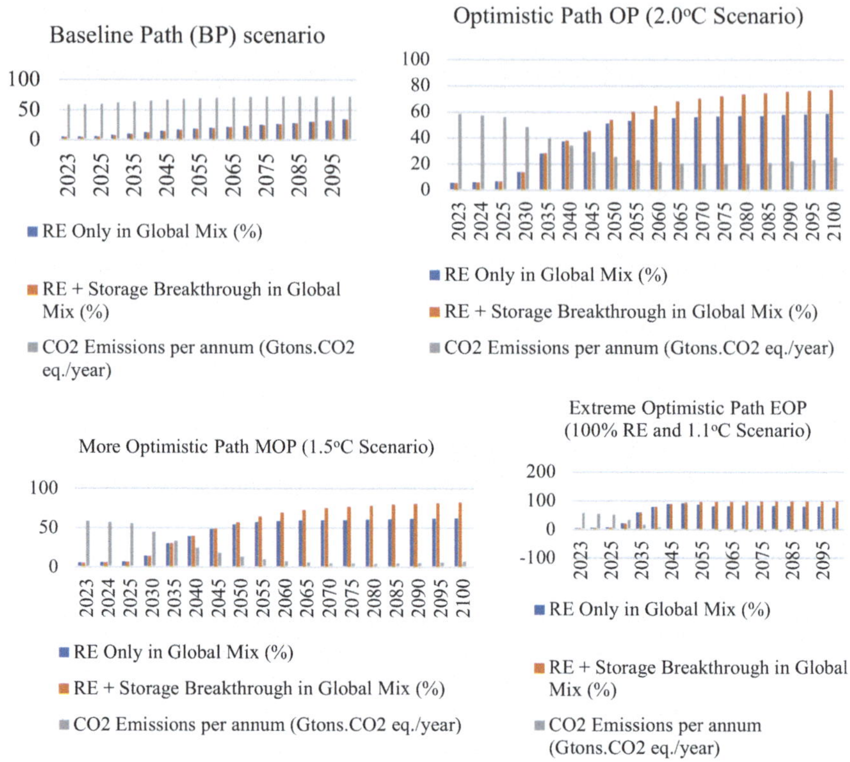

Fig. 3 Towards 100% RE versus temperature rise level: scenario results

the contribution from bioenergy, and with diminishing production from fossil-based fuels and non-renewable technologies such as coal, oil, and natural gas.

More importantly, the objective of this section is to discuss the renewable energy share, CO_2 emissions, and the average global temperature levels for the four scenarios. Hence, Fig. 3a–d depicts the summary of the values at a 5-year interval apart from the years 2023 and 2024, wherein the BP, OP, MOP, and EOP scenarios, the share of renewable energy production continues to increase. In the BP scenario, the share of storage breakthroughs is negligible; hence, RE share is almost the same as RE + storage, whereas in the OP, MOP, and EOP scenarios, storage capacity in the global mix increases annually and significant changes are observed from 2040. RE production in the global primary mix would reach 34.69, 77.3, 82.82, and approximately 100% for the BP, OP, MOP, and EOP, respectively, by 2100.

A unique result is seen in the EOP, where the global energy mix reaches 99% in 2070 and approximately 100% in 2090 (also shown in Table 3), but with a greater increase in the value of ratio storage breakthrough to RE production. The significance of this result shows that the promotion of energy storage is very instrumental to the 100% vision, as the resources for RE production are within

Table 3 Renewable energy production (% of total extra joules/year) comparison of the scenarios

Scenario	Temp (°C)	2030	2050	2070	2100
BP	3.32	**8.02** + *0* (644.72)	**16.79** + *0* (759.73)	**23.56** + *0* (872.99)	**34.69** + *0* (1072.14)
OP	2.0	**13.89** + *0.1* (524.31)	**51.05** + *3.18* (541.53)	**56.34** + *14.45* (673.29)	**58.57** + *18.74* (1057.26)
MOP	1.5	**14.12** + *0.1* (520.29)	**54.82** + *3.18* (540.24)	**60.01** + *15.7* (708.4)	**62.68** + *20.14* (1147.19)
EOP	1.1	**21.89** + *0.11* (411.32)	**91.60** + *3.29* (1360.09)	**84.45** + *14.74* (3604.54)	**77.51** + *22.49* (8850.54)

* Blue (% of total energy production from RE into the global primary mix), Green (extra joules, storage in the global primary mix), and Red (extra joules, energy for storage + total energy production or global primary production).

limits. These limits, such as biophysical and socio-economic constraints, have been discussed extensively in the MEDEAS WILLIAMS model of the LOCOMOTION projects of the European Union [23].

In terms of the CO_2 eq emissions, the BP scenario continues to see an increasing emission, with a peak at 72.19 Gtons CO_2 eq around 2085, a year where the world's population is expected to peak, and the developing countries may as well reach a high height of advancement in terms of 100% electrification and urbanisation (inclusive of industrial process, transport, and building). However, with the introduction of policies as in the OP, MOP, and EOP scenarios to support RE production, CO_2 emissions reduce across these three scenarios but rise again between 2070 and 2085 for all three scenarios.

The sudden rise is a result that the existing RE infrastructures are expected to exceed their useful life; hence, the extraction of material and development of newer infrastructures would be needed, giving rise to additional rise to CO_2 emissions. Such an occurrence could be avoided if there were technological breakthroughs in industrial processes that allow the production of RE infrastructures with a net-zero carbon footprint [24]. Most of these technologies are still in their infancy and have been highlighted in some studies [7, 25, 26].

Interestingly, in the EOP scenario, negative emissions are obtained, which is quite good for reaching a very safe climate scenario. However, it is important to see that this may pose a threat to other ecosystems, and if such a path is to be followed, measures are considered to address the impact of negative CO_2 emissions on the plant ecosystem and other important areas that use CO_2 as feedstocks.

The overall energy production (in extra joules) is expected to rise continually across all four scenarios, as shown in Fig. 4a. However, the increasing amount of production does not follow the same trajectory, as different decision choices have influenced this outcome. For instance, in the BP scenario, no storage breakthroughs are considered. Hence, high extra joules are expected due to the continued fossil fuel burning, which uses high-energy intensity to produce power compared with the OP scenario. Because of the high-energy intensity, the climate change potential of the CO_2 emitted is high, resulting in a 3.32 °C temperature, based on the model structure of the simulation tool.

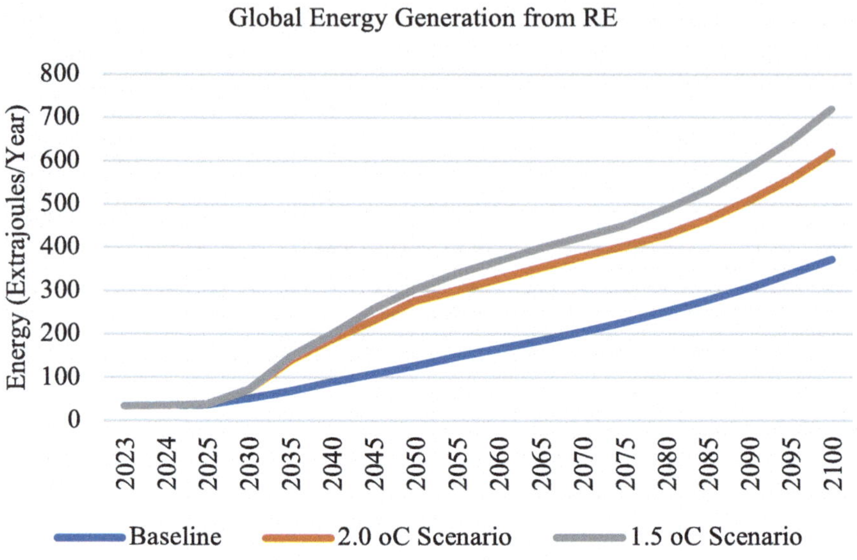

(a) Baseline, 2.0° C, and 1.5° C Scenarios

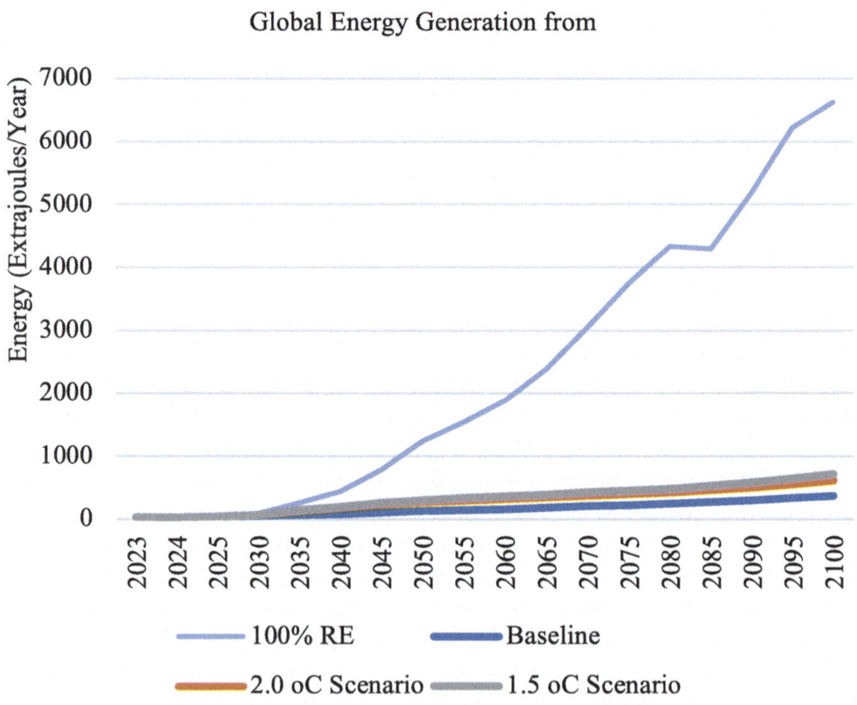

(b) Baseline, 2.0° C, 1.5° C, and 100% RE Scenarios

Fig. 4 Global Energy Generation from RE

Table 4 Emissions (Gton/eq.) comparison of the scenarios

Scenario	Temperature (°C)	2030	2050	2070	2100
BP	3.32	61.83	67.83	71.37	71.93
OP	2.0	48.09	25.5	19.73	24.82
MOP	1.5	44.55	12.95	4.68	7
EOP	1.1	33.19	−1.98	−7.08	−2.77

In the OP, MOP, and EOP, where storage is considered, energy production (in extra joules) increases with increasing storage capacity in the global primary mix. The increased extra joules can be attributed to the fact that additional electricity is needed to produce energy for storage. Since these extra joules would come from the high RE production share, the climate change potential of the energy intensity is reduced, resulting in the corresponding 2.0, 1.5, and 1.1 °C temperature levels for the OP, MOP, and EOP, respectively. By incorporating energy storage, a drop in carbon emissions through a decrease in continuous energy production can be achieved. However, such possibilities would come at a very high cost of tax to the population or government subsidy.

From additional findings, as summarised in Table 4 for the emission values at 2030, 2050, 2070, and 2100 across the four scenarios, the EOP, which is the only scenario with 100% RE possibilities, is not necessarily needed to keep the global temperature rise below 2 °C by 2100. Working towards a total of about 77.3% RE in the global energy mix is enough to keep the global temperature, even at 2 °C in support of the requirement for climate protection by the Paris Agreement. Similarly, at the same rate of 77.3% RE, the global carbon emissions would be dampened to 48.09, 25.5, 19.73, and 24.82 Gtons CO_2 eq in 2030, 2050, 2070, and 2100, respectively. However, this comes with a very high cost of carbon pricing while reducing the subsidies for coal, oil, natural gas, and bioenergy, as was specified in the scenario inputs from [12].

The choice of scenario has a significant impact on the global future energy mix, CO_2 emissions, and average global temperature levels. The BP scenario would lead to continued reliance on fossil fuels, high CO_2 emissions, and low-energy costs. The OP scenario would result in a cleaner energy mix, lower CO_2 emissions, and higher energy costs. The MOP scenario would provide a balance between environmental and economic considerations, with medium CO_2 emissions, medium energy costs, and a focus on energy efficiency as the world faces a critical decision about its future energy mix.

The BP scenario is unsustainable in the long run, as it would lead to high CO_2 emissions and global temperature rise beyond 2.0 °C. The OP and MOP scenarios are more aligned with the IPCC climate commitment recommendations. However, it would require significant investment in the renewable energy infrastructure of power generation to account for about 77.3% and 82.82% of the global primary production from RE, as in the OP and MOP scenarios, respectively.

The EOP scenario offers the best ground as it is in line with the 100% renewable energy vision for all purposes, but it would require a very strong commitment beyond just new renewable energy infrastructure. In addition, even though bioenergy is not categorised as renewable due to the release of some emissions, tax/subsidy on bioenergy had the least impact on the global temperature levels compared with fossil fuels. Hence, bioenergy may still be needed in the energy mix as its advantages and environmental friendliness compared to fossil fuel cannot be overemphasised. These options need to be carefully considered in developing a comprehensive energy policy that balances all or most of the major approaches, strategies, and considerations made in the simulation.

Based on the results, a huge difference is seen in the extra joules needed for the 100% RE scenario (Fig. 4b), which is capable of keeping the global temperature levels at 1.1 °C. For instance, about 6627 extra joules are needed in the 100% RE scenario, compared with the 719 and 619 extra joules for the 2.0, 1.5 °C scenarios. Such a possibility would require unprecedented supporting actions to support such to happen, as well as dedicated efforts to achieve a high share of energy storage in the primary energy mix. Interestingly, these large extra joules can be drastically reduced if research breakthroughs result in a very high reduction of electricity requirement for energy storage production, such as hydrogen.

5 Limitation and Conclusion

This study examines the feasibility of implementing a global 100% renewable energy system in comparison to long-term climate change scenarios using the EN-ROADS integrated assessment framework. Four different energy policies are hypothesised for the years 2023–2100: baseline path (BP), optimistic path (OP), more optimistic path (MOP), and extreme optimistic path (EOP) policies. The study aims to bridge gaps between existing studies and those on developing 100% renewable energy global systems. The simulation model of EN-ROADS is used to investigate the feasibility of a 100% renewable energy mix based on global exogenous and endogenous assumptions. Four scenarios are developed, including baseline path (BP), optimistic path (OP), more optimistic path (MOP), and extreme optimistic path (EOP) policy scenarios in a single EN-ROADS model. These scenarios compare both renewable energy and fossil fuel targets with other sectors, growth, emissions, and emission removal technologies. The precise influence on global temperature predictions depends on technological advancements, global collaboration, and the promptness of implementation. Adopting a complete reliance on renewable energy sources presents a hopeful approach to attaining the worldwide temperature objectives specified in the Paris Agreement. However, the study has limitations, such as not prioritising different renewable energy sources, not considering the potential of individual renewable energies, and not evaluating the risks associated with investments. In general, this study research analyses the relationship between the transition to 100% renewable energy and the attainment of lower global temperatures. The success of this shift depends on the promptness of implementation, technological progress, and global cooperation.

Acknowledgement The authors appreciate the support of the Durban University of Technology under the Postgraduate RFA-Energy research scholarship.

Conflict of Interest and Data Statement The authors declare no conflict of interest, and the data utilised for this study is available upon request.

References

1. Liu, J.Y., et al.: The importance of socioeconomic conditions in mitigating climate change impacts and achieving Sustainable Development Goals. Environ. Res. Lett. **16**(1) (2020). https://doi.org/10.1088/1748-9326/abcac4
2. Olabi, A.G., Abdelkareem, M.A.: Renewable energy and climate change. Renew. Sust. Energ. Rev. **158** (2022, April). https://doi.org/10.1016/j.rser.2022.112111
3. Akpan, J., Olanrewaju, O.: Towards the 1.5°C climate scenario: global emissions reduction commitment simulation and the way forward. In: Global Warming – A Concerning Component of Climate Change [Working Title], 1st edn. IntechOpen, London (2023). https://doi.org/10.5772/intechopen.1003851
4. Taconet, N., Méjean, A., Guivarch, C.: Influence of climate change impacts and mitigation costs on inequality between countries. Clim. Chang. **160**(1), 15–34 (2020, May). https://doi.org/10.1007/s10584-019-02637-w
5. European Environment Agency and European Commission. Climate Action DG: Climate-ADAPT: 10 Case Studies: How Europe is Adapting to Climate Change. Publications Office of the European Union (2018)
6. Aniebo, I.N., Akpan, J.S.: Energy transition: implications, considerations, and roadmap development for Sub-Saharan Africa. In: *SPE Nigeria Annual International Conference and Exhibition*, OnePetro. 2022, August, pp. 1–10. https://doi.org/10.2118/211990-MS
7. Akpan, J., Olanrewaju, O.: Sustainable energy development: history and recent advances. Energies (Basel). **16**(20), 7049 (2023, October). https://doi.org/10.3390/en16207049
8. IPCC: AR5 Synthesis Report: Climate Change 2014The Synthesis Report (SYR) of the IPCC Fifth Assessment Report (AR5), October 2014
9. Asmelash, E., Prakash, G., Gorini, R., Gielen, D.: Role of IRENA for Global Transition to 100% Renewable Energy, 2020, pp. 51–71. https://doi.org/10.1007/978-3-030-40738-4_2
10. Akpan, J., Olanrewaju, O.: Towards a common methodology and modelling tool for 100% renewable energy analysis: a review. Energies (Basel). **16**(18), 6598 (2023, September). https://doi.org/10.3390/en16186598
11. Siegel, L.S., Campbell, C., Fiddaman, T., Jones, A.P., Jones, C., Sterman, J.: En-ROADS Technical Reference, August 2023. Accessed 22 Aug 2023. [Online]. https://docs.climateinteractive.org/projects/en-roads-reference-guide/en/latest/ref-guide.pdf
12. Akpan, J.: Development of Integrated Model and Framework for Sustainable Energy Resources and Systems Planning. Durban University of Technology, Durban (2024)
13. IEA: World Energy Outlook 2022, Paris, October 2022. Accessed 14 Aug 2023. [Online]. https://www.iea.org/reports/world-energy-outlook-2022
14. British Petroleum: Statistical Review of World Energy 2022, June 2023. Accessed 28 July 2023. [Online]. https://www.energyinst.org/statistical-review
15. IRENA: Perspectives for the Energy Transition- Investment Needs for a Low-Carbon Energy System, 2017. Accessed 19 Sep 2023. [Online]. https://www.irena.org/-/media/Files/IRENA/Agency/Publication/2017/Mar/Perspectives_for_the_Energy_Transition_2017.pdf?rev=46ff9f7aebec4cd7a983fd8b13697648

16. Millstein, D., Wiser, R., Bolinger, M., Barbose, G.: The climate and air-quality benefits of wind and solar power in the United States. Nat. Energy. **2**(9), 17134 (2017, August). https://doi.org/10.1038/nenergy.2017.134
17. Kyne, D., Bolin, B.: Emerging environmental justice issues in nuclear power and radioactive contamination. Int. J. Environ. Res. Public Health. **13**(7), 700 (2016, July). https://doi.org/10.3390/ijerph13070700
18. Dolphin, G.: World Carbon Pricing Database: Resources for the Future. https://www.rff.org/publications/data-tools/world-carbon-pricing-database/. Resources for the Future (2022)
19. Rosenow, J., Eyre, N., Sorrell, S., Guertler, P.: The EU Referendum: Implications for UK Energy Policy Briefing Unlocking Britain's First Fuel: The Potential for Energy Savings in UK Housing Key Findings, 2017. Accessed 19 Aug 2023. [Online]. http://e3g.wpengine.com/wp-content/uploads/docs/17_9_6_UKERC_CIED_briefing_final.pdf
20. Rojas-Rueda, D., de Nazelle, A., Teixidó, O., Nieuwenhuijsen, M.J.: Replacing car trips by increasing bike and public transport in the greater Barcelona metropolitan area: A health impact assessment study. Environ. Int. **49**, 100–109 (2012, November). https://doi.org/10.1016/j.envint.2012.08.009
21. Royal Society (Great Britain) and Royal Academy of Engineering (Great Britain): *Greenhouse Gas Removal*. 2018. Accessed 17 Sep 2023. [Online]. https://royalsociety.org/-/media/policy/projects/greenhouse-gas-removal/royal-society-greenhouse-gas-removal-report-2018.pdf
22. United Nations: 2022 Revision of World Population Prospects. United Nations. https://population.un.org/wpp/ (2022)
23. Capellán-Pérez, I., et al.: MEDEAS: a new modeling framework integrating global biophysical and socioeconomic constraints. Energy Environ. Sci. **13**(3), 986–1017 (2020). https://doi.org/10.1039/C9EE02627D
24. Akpan, J., Olanrewaju, O., Irusta, R.: A state-of-the-art approach for assessing the environmental sustainability of multi-renewable energy systems in the built environment. In: *Advances in Clean Energy Systems and Technologies*, Green Energy and Technology-Springer Book Series, Springer, Cham (2024). https://doi.org/10.1007/978-3-031-49787-2_31
25. IEA: Energy Technology Perspectives 2023, 2023. [Online]. www.iea.org
26. Calise, F.: Recent advances in green hydrogen technology. Energies. **15**(16) (2022, August 1). https://doi.org/10.3390/en15165828

Part III
Energy Materials and Energy Storage Technology

Research on the Performance of Positive Electrode Modified with Fluorinated Carbon Based on Silver Coating

Sun Haitao, Cui Kaibo, Deng Huiyong, and Wei Zhenjie

1 Introduction

The research on fluorination process for carbon materials is now more mature and can meet the demand of commercial production [1, 2]. However, the balance between CFx energy density and power density still needs to be further optimized. Coating is the best way to solve this contradiction, so different substances such as alloys, oxides, and carbides have been used as coating materials and have improved the power density and cycle stability of CFx to a certain extent. However, there is still a lack of in-depth understanding of the coating mechanism and the enhancement strategy of the improvement of the effect of these different coating materials [3, 4]. This paper mainly focuses on the in-depth study of Ag cladding materials for CFx.

Common types of metal conductivity are different, [5, 6] usually the best conductivity of Ag, followed by Cu and Au. Ag's physicochemical properties are more stable, thermal conductivity, electrical conductivity is very good, the activity is weaker than hydrogen, in general, cannot be reacted with dilute acid replacement H_2, so it is very difficult to react with the HF in the electrolyte, the materials play a very good role in protection. The researchers prepared Ag-coated Ni particles by ball milling and thermal plasma process, and the morphological and structural analysis showed that the synthesized material has reticularly connected Ag channels and dispersed Ni cores, in which the Ag channels can act as highly conductive paths and improve the physical properties of the material. Recently, researchers used co-precipitation to encapsulate Ag in NCM811, a cathode material for lithium-ion batteries, which effectively improved the electrochemical performance of NCM811. Therefore, in this paper, the technical route of co-precipitation of $AgNO_3$ and

S. Haitao (✉) · C. Kaibo · D. Huiyong · W. Zhenjie
Shijiazhuang Campus of Army Engineering University, Shijiazhuang, China
e-mail: sunhaitao@aeu.edu.cn

L. Chen (ed.), *Trends in Clean Energy Research*, Green Energy and Technology,
https://doi.org/10.1007/978-3-031-67987-2_16

$NH_3 \cdot H_2O$ + high temperature calcination at 550 °C is proposed to be used to prepare Ag-coated fluorinated carbon materials. By studying the mechanism of Ag modification on the electrochemical performance of fluorocarbon, the effect of Ag coating on the electrochemical performance of fluorocarbon was analyzed to determine the best technical route.

2 Experimental Section

In this research, we propose to prepare Ag-coated fluorocarbon materials by the technical route of co-precipitation of $AgNO_3$ and $NH_3 \cdot H_2O$ high-temperature calcination at +550 °C.

The synthesis mechanism of the Ag cladding layer can be obtained according to the following steps:

$$AgNO_3 + NH_3 \cdot H_2O = NH_4 \cdot NO_3 + AgOH \downarrow \tag{1}$$

$$2AgOH = H_2O + Ag_2O \tag{2}$$

$$2Ag_2O = 4Ag + O_2 \uparrow \tag{3}$$

First, 1 g of fluorinated graphite powder was homogeneously dispersed in an anhydrous ethanol solution at 50 °C, and then a certain mass fraction of $AgNO_3$ powder was homogeneously dissolved in an anhydrous ethanol solution at 50 °C. After ultrasonically treating both materials for 2 h, the $AgNO_3$ ethanol solution was added dropwise gradually to the fluorinated carbon-in-ethanol solution at a feed rate of 0.3 $mL \cdot min^{-1}$. At the same time, the above mixture was stirred continuously at 50 °C at 450 $r \cdot min^{-1}$, and then an appropriate amount of $NH_3 \cdot H_2O$ was added to the above mixture, controlling the pH of the solution to be 10.5. The anhydrous ethanol was completely volatilized by nonstop heating and stirring of the mixture, and then it was dried under vacuum for 5 h at 130 °C. Finally, the dried mixture was calcined at 550 °C in an air atmosphere for 5 h. After cooling naturally to room temperature, the material was removed and ground with a mortar and pestle for 1 h. The material was then sieved through a 200-mesh sieve to form Ag-coated co-modified fluorinated graphite material with uniform particle size. For comparison, the materials with 1 *wt%*, 2 *wt%*, and 3 *wt%* Ag coating are labeled as Ag-1, Ag-2, and Ag-3, respectively.

3 Ag-Coated Fluorocarbon Modification Study

3.1 Effect of Ag Coating on the Composition and Elemental Distribution of Fluorocarbon Phases

In order to investigate the existence of Ag phase in the composites, the XRD spectra of the pristine fluorinated graphite and different ratios of Ag-coated materials are given in Fig. 1.

As can be seen from the figure, the pristine fluorinated graphite exhibits a typical broad peak belonging to carbon at around 25°. In addition, the intensity of the peaks at (111), (200), (220), and (311) increases with the addition of Ag, indicating the increase of silver content. Moreover, these peaks are present in the same position in all the coated samples, implying that the introduced Ag does not affect the basic structure of fluorinated graphite. From the standard card JCDPS87–0697, the lattice parameter for the introduction of Ag is a = b = c = 4.086. However, when the amount of Ag coating reaches 3 *wt%*, the peak intensities of (111), (200), (220), and (311) increase dramatically, indicating that an excessive amount of Ag has been introduced into the carbon matrix.

SEM can be used to directly image the material properties of the sample surface material microscopically. SEM is, in principle, the use of very finely focused high-energy electron beams scanned on the specimen and excitation of a variety of physical information. By accepting, magnifying, and displaying the imaging of this information, an observation of the surface topography of the test specimen is obtained.

As shown in Fig. 2a, the pristine fluorinated graphite exhibits a clear lamellar structure, and these morphologies are still well preserved even with the introduction of the Ag capping agent (Fig. 2c, d). It is also important to note that inside these samples of Ag-1/Ag-2/Ag-3, the layer structure becomes denser and therefore a

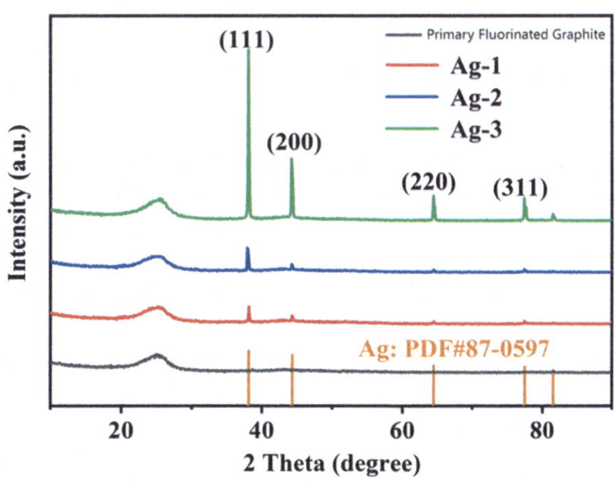

Fig. 1 XRD for different samples

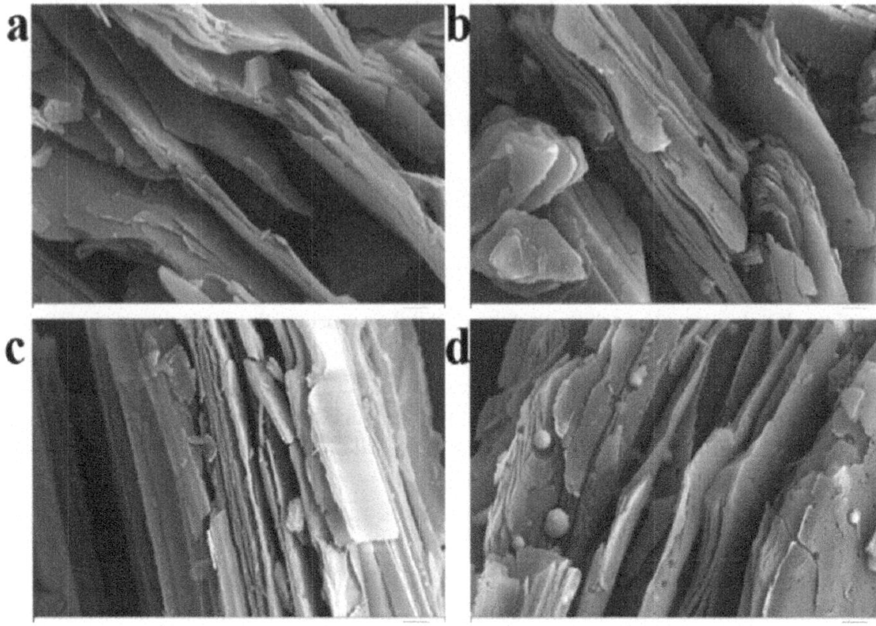

Fig. 2 SEM for different samples. (**a**) Original sample, (**b**) Ag-1, (**c**) Ag-2, and (**d**) Ag-3

higher number of layers are formed. The overly dense structure, however, is not only unfavorable for ion transfer, but also the volumetric strain during charge storage can cause severe structural fragmentation, thus resulting in rapid capacity decay. In addition, we picked Ag-2 samples for elemental distribution tests (Fig. 3), which showed that Ag elements were uniformly distributed into the carbon matrix.

3.2 *Effect of Ag Coating on the Bonding Mode of Fluorinated Carbon*

As shown in Fig. 4, IR spectroscopy tests indicate that a small amount of Ag incorporation does not cause significant peak changes, e.g., pristine fluorinated graphite and Ag-1 exhibit similar characteristic peaks. When the addition of Ag is 2 *wt%*, the intensities of the characteristic peaks located at 1209.1 and 2354.6 cm^{-1} are obviously strengthened, with the former belonging to the vibration of the C–C bond and the latter related to the stretching vibration of $C \equiv C$ in fluorinated carbon. The vibrational peaks of C–C and $C \equiv C$ are further enhanced with the further increase of Ag incorporation. These results suggest that the introduced Ag cladding is improving the charge storage process by changing the bonding form of

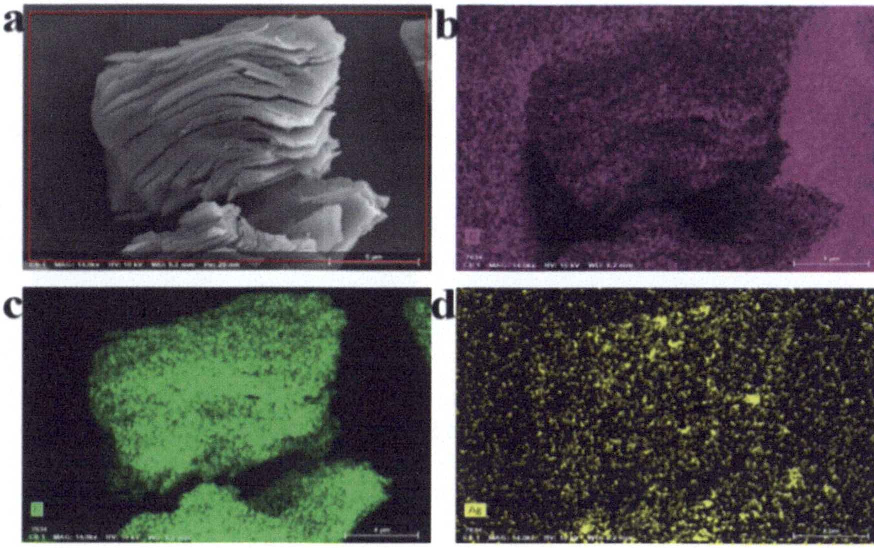

Fig. 3 Elemental distribution of Ag-2 samples

Fig. 4 Infrared test spectra of different samples

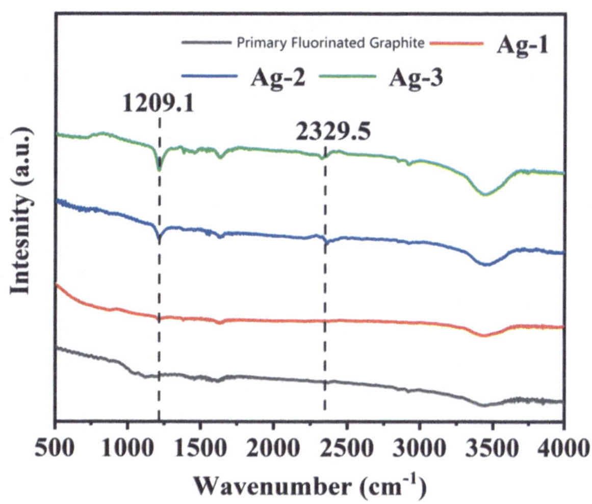

the fluorocarbon matrix. Therefore, in order to realize the optimal electrochemical performance, it is necessary to rationally control the introduction amount of Ag.

Figure 5a shows the C 1 s XPS spectral peaks of pristine fluorinated graphite and Ag-1/Ag-2/Ag-3 samples, which mainly include C–C, C–O, and O–C=O functional groups. And the positions of these peaks did not shift with the addition of Ag, indicating that the bonding form of the carbon matrix remained stable. As shown in Fig. 5b, the F 1 s peak of the pristine fluorocarbon is mainly composed of CF2 bonds, and the position of this peak did not change when the addition of Ag was

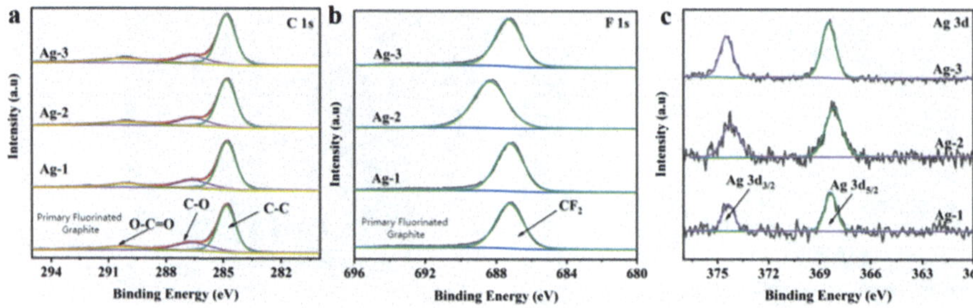

Fig. 5 XPS test spectra for different samples. (**a**) C 1 s, (**b**) F 1 s, and (**c**) Ag 3d

1 *wt%*. With the increase of Ag addition to 2 *wt%*, the CF2 bond is shifted to a high binding energy position, implying a change in the bonding form of the F element. However, for the Ag-3 sample, the CF2 bond position remains the same as that of pristine fluorinated graphite, suggesting that too much Ag content may not have a gain effect on the electrochemical properties. As can be seen in Fig. 5c, the strength of Ag-2$d_{3/2}$ as well as Ag-2$d_{5/2}$ is subsequently strengthened with the increase of Ag content.

3.3 *Effect of Ag Coating on the Electrochemical Properties of Fluorocarbons*

Figure 6 shows the discharge curves of the raw fluorinated graphite and the three modified materials under different multiplicity conditions at 20 °C in turn. As shown in Fig. 6a, the discharge capacities of the raw fluorinated graphite at 0.1C, 0.2C, 0.5C, 0.8C, and 1C (1C = 809.3 mAh/g) were 886.4, 656.8, 552.2, 487.9, and 349.2 mAh/g, respectively, and the discharge curves had a clear plateau at each multiplication rate. However, with the introduction of Ag capping agent (Ag-1 sample), the voltage plateau at each multiplicity diminished (Fig. 6b), although the discharge capacity did not change significantly. When the addition of Ag was increased to 2 *wt%* (Fig. 6c), the discharge plateau disappeared, but the discharge capacities at 0.1C, 0.2C, 0.5C, 0.8C, and 1C were 959.5, 720.9, 621.2, 511.7, and 409.7 mAh/g, respectively. The Ag-3 sample (Fig. 6d) had a similar charge/discharge profile to that of Ag-2. However, its capacity is much lower than that of pristine fluorinated graphite. These results indicate that the introduction of Ag cladding can affect the discharge behavior of fluorinated graphite, and the electrochemical performance is optimal when the addition is 2 *wt%*.

As shown in Fig. 7a, the discharge median voltage of pristine fluorinated graphite decreases with increasing discharge multiplicity. For the Ag-1 sample, the trend of the median voltage evolution is similar to that of pristine fluorinated graphite, implying that the small amount of introduced Ag does not have a significant effect

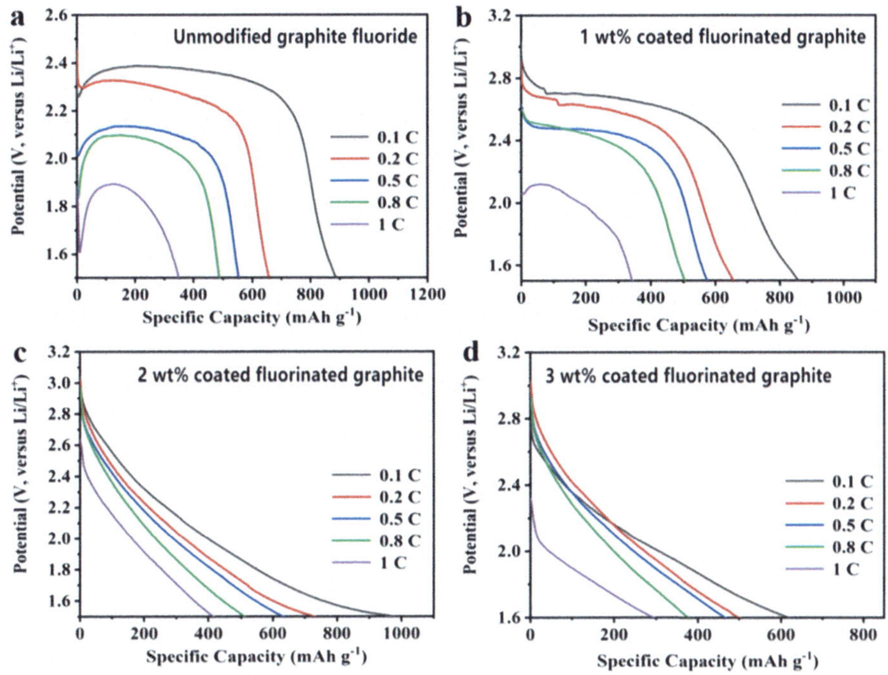

Fig. 6 Discharge diagram of different multiplicities at 20 °C. (**a**) Original sample, (**b**) Ag-1, (**c**) Ag-2, and (**d**) Ag-3

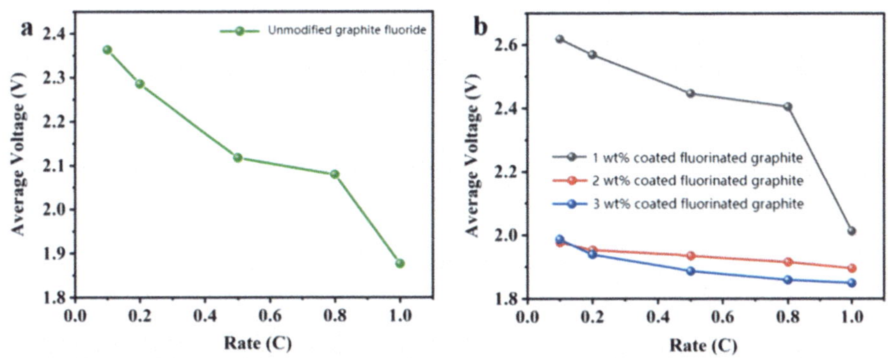

Fig. 7 Capacity graph at different magnifications at 20 °C. (**a**) Original sample, (**b**) Ag-1

on the discharge behavior. However, both Ag-2 and Ag-3 samples exhibit low median voltages (Fig. 7b), with values of just 1.97 V and 1.98 V at a multiplicity of 0.1C in time, suggesting an alteration of the discharge mechanism that corresponds well to their respective discharge curves.

4 Summary

The uniform coating of Ag on the surface of CFx can be realized by co-precipitation and high-temperature calcination processes. The coated Ag particles will combine with F in CFx to form AgFy, which in turn reduces the F content in CFx. A small amount of Ag coating (1 *wt%*) leads to a decrease in the specific capacity, an increase in the electronic conductivity, and the multiplicity performance of CFx, while too much Ag coating (2 *wt%*) leads to the similarity of the discharge characteristics of CFx with graphite and a significant decrease in the median voltage of the discharge. The maximum discharge-specific capacity of 959.5 mAh-g-1 was achieved when the addition of Ag was 2 *wt%*, while the different ratios of Ag-doped CF materials were able to achieve discharge in a wide temperature region from -40 to 40 °C.

References

1. Xu, G., Yuan, J., Tao, X., et al.: Absorption mechanism of carbon-nanotube paper-titaniumdioxide as a multifunctional barrier material for lithium-sulfur batteries[J]. Nano Res. **8**(9), 3066–3074 (2015)
2. Zhang, Q., Takeuchi, K.J., Takeuchi, E.S., et al.: Progress toward high-power Li/CFx batteries: electrode architectures using carbon nanotubes with CFx[J]. Phys. Chem. Chem. Phys. **17**(35), 22504–22518 (2015)
3. Liu, Z., Su, H., Yang, Y., et al.: Advances and perspectives on transitional metal layered oxides for potassium-ion battery[J]. Energy Storage Mater. **34**, 211–228 (2021)
4. Xu, L., Xiong, P., Zeng, L., et al.: Facile fabrication of a vanadium nitride/carbon fiber composite for half/full sodium-ion and potassium-ion batteries with long-term cycling performance [J]. Nanoscale. **12**(19), 10693–10702 (2020)
5. Jiang, C., Li, X., Ying, Y., et al.: Fluorinated graphene-enabled durable triboelectric coating for water energy harvesting [J]. Small. **17**(8), e2007805 (2021)
6. Jones, J.-P., Jones, S.C., Billings, K.J., et al.: Radiation effects on lithium CFx batteries for future spacecraft and landers [J]. J. Power Sources. **471**, 228464 (2020)

Research on the Positive Electrode Performance of Lithium Battery Based on the Mixture of Carbon Fluoride and Manganese Dioxide

Cui Kaibo, Sun Haitao, Deng Huiyong, and Weng Huihuang

1 Experimental Part

Lithium-fluorocarbon (Li-CFx) batteries have the largest theoretical capacity compared to other lithium primary batteries, such as Li-$SOCl_2$ and Li-MnO_2 batteries [1, 2]. In addition, Li-CFx battery has long storage life, stable operation, and wide operating temperature range. However, due to the strong covalency of the C–F bond, fluorocarbons exhibit very low electronic conductivity and are prone to polarization, which leads to the fast attenuation of battery performance [3, 4]. In order to solve this problem, the study of α-MnO_2 and fluorocarbon composite was carried out in this paper, in order to further improve the magnification properties of fluorocarbon cathode materials and explore the influence of different mass ratios of α-MnO2 on the electrochemical properties of composite materials.

1.1 *Synthesis of CFx/MnO_2 by Solid-Phase Method*

A planetary ball mill is used in this experiment. The following is a brief introduction to the two materials:

1. In this experiment, fluorocarbons and α-MnO_2 were used as raw materials. In order to explore the effect of the ratio of CFx/MnO_2 synthesis, the molar ratios were 9/1, 8/2, and 7/3.
2. Surface dispersant. Because the surface energy of fluorocarbon is relatively low, it is difficult to dissolve in water, and a certain amount of surfactant must be

C. Kaibo (✉) · S. Haitao · D. Huiyong · W. Huihuang
Shijiazhuang Campus of Army Engineering University, Shijiazhuang, China
e-mail: cuikb@aeu.edu.cn

L. Chen (ed.), *Trends in Clean Energy Research*, Green Energy and Technology,
https://doi.org/10.1007/978-3-031-67987-2_17

added to reduce its surface tension significantly, so that the particle surface can get a good wetting effect. In the experiment of this paper, polyether P-123 was used as the dispersing agent to realize the dispersion of CFx in water.
3. Bottom liquid. The bottom liquid is anhydrous ethanol.

1.2 Specific Experimental Operation Steps

1. Place the planetary ball mill in a flat and open position on the ground, maintain balance, open the instrument idling test run for 3–5 min, observe whether there is a problem with the instrument panel, and determine whether there is no noise inside the machine during operation.
2. After the trial operation is corrected, the ball grinding tank and the ball are cleaned.
3. Put the prepared fluorocarbon and α-MnO_2 samples and the grinding ball into the ball grinding tank in proportion, and then add anhydrous ethanol, paying attention to reserving one-third of the grinding space.
4. Open the lid, fix the ball mill tank with the ball material in the tank seat, and close the lid.
5. The equipment is powered on, turn on the switch and set the low speed of 100 r/min to run for 10 s, and then adjust to the high speed of 500 r/min ball milling for 2 h after no abnormality is observed.
6. Put the ground solution into a drying oven and dry it at a temperature of 80 °C for 6 h to obtain the precursor material.

2 Study on the Properties of Mixed Positive Electrode of Fluorocarbon/Manganese Dioxide

2.1 Structural Characterization of Mixed Positive Electrode of Fluorocarbon/Manganese Dioxide

Figure 1 shows the SEM microstructure characterization of pure CFx and different proportions of CFx/MnO_2 composites. The pure CFx in Fig. 1a shows the micromorphology of multilayer accumulation, which is conducive to the full embedding of lithium ions. Figure 1b–d shows the morphological characterization of CFx mixed with MnO_2 in different proportions. It can be obviously seen that with the increase of MnO_2 incorporation ratio, the local distribution of rod-shaped MnO_2 increases significantly, and when the ratio of CFx/MnO_2 reaches 7/3, it even shows a phenomenon of serious MnO_2 enrichment. However, when the content of MnO_2 in Fig. 1b is small, the overall doping effect is not obvious. When the doping ratio reaches 8/2, the distribution of MnO_2 shows a relatively uniform effect. MnO_2

Fig. 1 SEM images with different MnO_2 doping ratios. (**a**) pure CFx, (**b**) 9/1, (**c**) 8/2, and (**d**) 7/3

distributed among layered CFx can provide transmission channels for electrons and ions to a certain extent and promote the improvement of the discharge rate and discharge capacity.

The above SEM results show that the best doping effect is obtained when the MnO_2 ratio is 8/2. As shown in Fig. 2, the regional element distribution energy spectrum test was carried out on the SEM image at this ratio. Through the analysis of the type and content of elements in the microregion of the material, the height correspondence of element mapping showed that the composite material was composed of rodlike MnO_2 and lamellar CFx.

2.2 *Electrochemical Properties of Fluorocarbon/Manganese Dioxide Mixed Positive Electrode*

Figure 3 shows the rate discharge performance test diagram of CFx/MnO_2 composites with different mixing proportions at 20 °C, and the selected rate gradients are 0.1C, 0.5C, 1C, 2C, and 5C. Since both components contribute capacity during discharge, they can be used as active substances for calculation. The theoretical specific capacity of CFx is 865 mAh·g^{-1} and that of α-MnO_2 is 308 mAh·g^{-1} according to the mass ratio. By calculation, the theoretical specific capacity of 9/1 mass ratio is 809.3 mAh·g^{-1}, the theoretical specific capacity of 8/2 mass

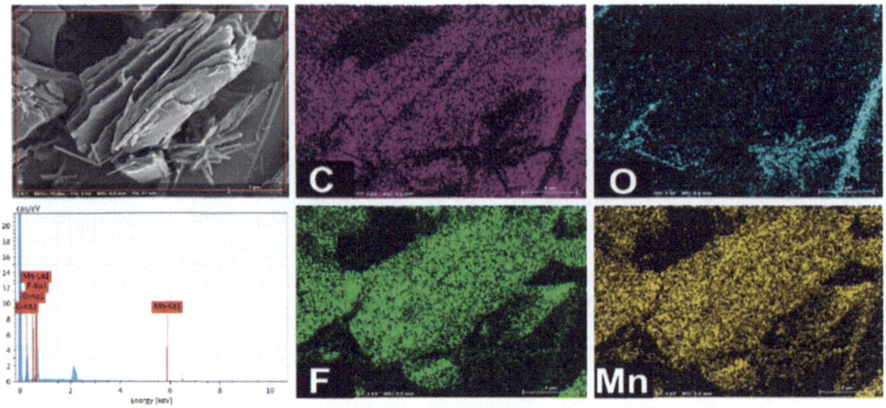

Fig. 2 Energy spectrum analysis of elements with 8/2 MnO_2 doping ratio

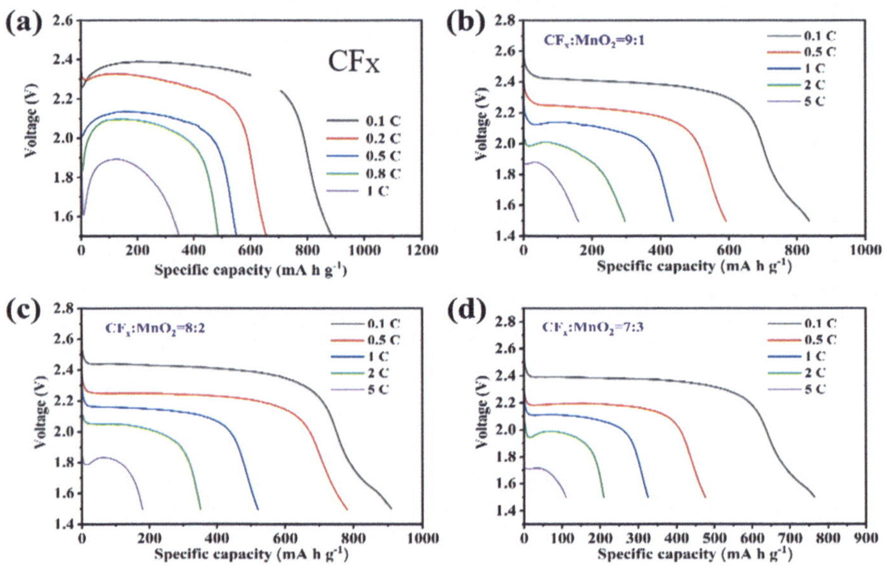

Fig. 3 Rate discharge diagram of CFx/MnO_2 with different mass ratios at 20 °C. (**a**) pure CFx, (**b**) 9/1, (**c**) 8/2, and (**d**) 7/3

ratio is 753.6 mAh·g^{-1}, and the theoretical specific capacity of 7/3 mass ratio is 697.9 mAh·g^{-1}. The discharge current (C) at different rates is determined by the theoretical specific capacity calculation of each component, and the discharge cutoff voltage is 1.5 V.

As shown in Fig. 3, the sample with a mass ratio of 8/2 has the best specific capacity performance at all current rates and has a specific capacity of 909.85 mAh·g^{-1} at a low current density of 0.1C, which is higher than that of the original

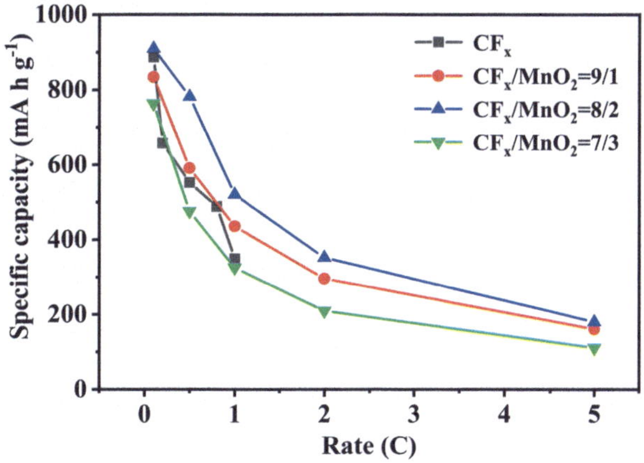

Fig. 4 Magnification-specific capacity diagram of unmodified CFx and different mass ratios of CFx/MnO_2 at 20 °C

CFx sample. However, when the incorporation of α-MnO_2 was higher than and lower than 8/2, the capacity of 834.88 mAh·g^{-1} and 763.64 mAh g^{-1} (at 0.1C) was lower than that of pure CFx samples, indicating that the introduction of appropriate proportion of MnO_2 had a significant effect on the capacity improvement at a small magnification rate. Too much or too little MnO_2 content will hinder the play of the intrinsic capacity. When the rate current reaches 2C, only the sample with the ratio of 8/2 can maintain a relatively stable discharge platform voltage, while the other samples have voltage return phenomenon caused by polarization. When the magnification current reaches 5C, all the samples show insignificant platform capacity.

In addition, the rate-specific capacity plot in Fig. 4 provides a more detailed comparison of the discharge capacity differences between MnO_2 and the unmodified pure CFx. It can be seen that except for the electrochemical performance when the incorporation ratio is 8/2, the capacity of other incorporation ratios is lower than that of the original sample, indicating that proper incorporation of MnO_2 can significantly improve the discharge effect of the battery.

The median voltage of the three proportional samples under different rate currents is shown in Fig. 5. When the mixing ratio is 8/2, the platform voltage is the highest at each ratio, and the discharge capacity is superior to other ratios. In addition, at low rate current density (0.1C and 0.2C), the three samples with different proportions show a small voltage platform gap, and with the further increase of current density, especially when increasing to 2C, the discharge platform of the sample with the proportion of 8/2 has a significant gap with the other two samples. At this time, the samples containing more MnO_2 (8/2 and 7/3) showed a higher discharge platform, indicating that MnO_2 had an obvious effect on the increase of discharge platform voltage at the intermediate rate discharge current. However, as the magnification current increased to 5C, the medium voltage of the samples with less MnO_2 content (9/1 and 8/2) showed a high-voltage advantage, and the difference between the samples with 7/3 ratio was obvious, indicating that the high

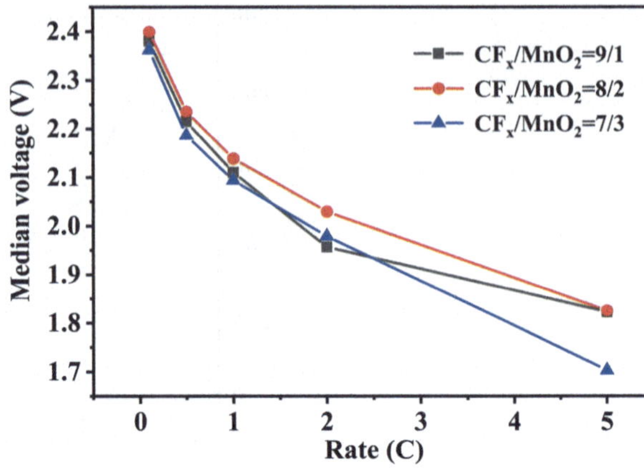

Fig. 5 Median voltage diagram of CFx/MnO_2 with different mass proportions at different magnification at 20 °C

content of CFx had a positive effect on the enhancement of the high-voltage platform at the large magnification current.

2.3 AC Impedance Analysis of Fluorocarbon/Manganese Dioxide Mixed Positive Electrode

In order to study the effect of different contents of MnO_2 on the electrochemical performance of composite materials, AC impedance test was used to study the composite positive electrode with different contents of MnO_2, and Z view software was used to fit the measured impedance curve [5, 6]. The test requires the assembled half battery to stand for 10 h and to be carried out in a stable voltage state after the electrolyte fully contacts the electrode.

Figure 6 shows the AC impedance diagram of prepared CFx/MnO_2 composites of different proportions. This electrochemical impedance result is fitted by the equivalent circuit in Fig. 7, where Rs and Rct represent solution resistance and charge transfer resistance, respectively. The size of Rs is related to the resistance between the working electrode and the opposite electrode in the solution, including the combination of the collector, the electrolyte, the diaphragm, and the electrode. Rct occurs during the transfer from a solid (electrode) to a liquid (electrolyte), and its size can be measured by the diameter of the middle and high-frequency region in the impedance spectrum. According to equivalent circuit simulation, it is found that the simulated impedance curve is in good agreement with the measured impedance, and the specific simulation values are shown in Table 1.

As can be seen from Table 1, the Rct values of CFx/MnO_2 of the three ratios are 603.85, 704.21, and 566.42 Ω, respectively. With the increase of MnO_2 incorporation ratio, Rct showed first a decreasing trend and then an increasing trend. The early increase process is because MnO_2 has good electrical conductivity and can reduce the charge transfer impedance between particles, so that the impedance value

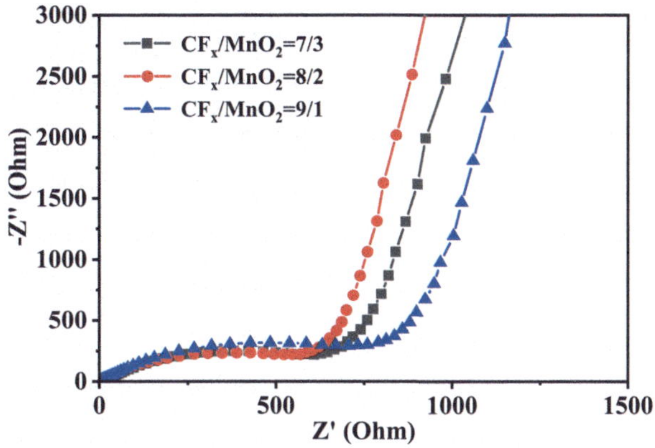

Fig. 6 AC impedance diagram of CFx/MnO_2 composites with different contents

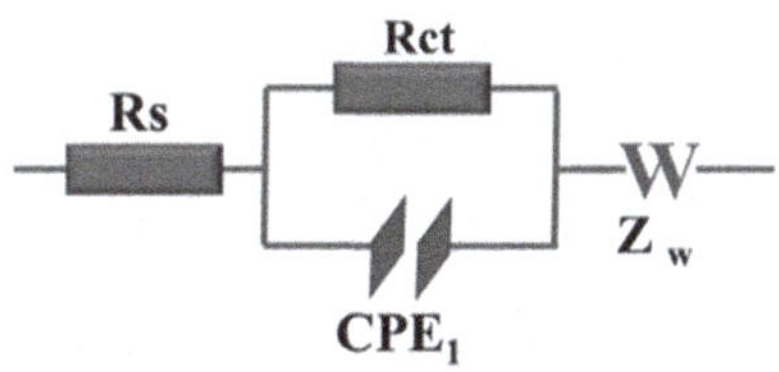

Fig. 7 Impedance analog equivalent circuit diagram of CFx/MnO_2 composites with different contents

Table 1 Impedance simulation values of CFx/MnO_2 composites with different contents

Sample	$R_s(\Omega)$	$R_{ct}(\Omega)$
$CF_x/MnO_2 = 7/3$	4.93	603.85
$CF_x/MnO_2 = 8/2$	2.45	566.42
$CF_x/MnO_2 = 9/1$	1.62	704.21

is significantly reduced. However, when the content of MnO_2 increases further, the charge transfer impedance also increases further, because too much MnO_2 will hinder the deep diffusion process of Li^+. Among them, when the ratio is 8/2, the impedance value is the smallest, indicating the best conductivity and discharge performance, which is consistent with the above discharge results.

The line slope in the low-frequency region is related to the solid-phase diffusion of Li^+ within the active substance, as shown in Fig. 8. The parameters of the diffusion process can be further fitted through the fitting relationship between Z' and $\omega^{-1/2}$. For specific calculations, see formula (1) and formula (2). The CFx/MnO_2 ratio of 8/2 shows a small D value (124.21), which has the fastest Li^+ diffusion efficiency:

$$D = \frac{R^2T^2}{2A^2n^4F^4C^2\sigma_w^2} \tag{1}$$

$$Z' = R_e + R_{ct} + \sigma_w\omega^{-1/2} \tag{2}$$

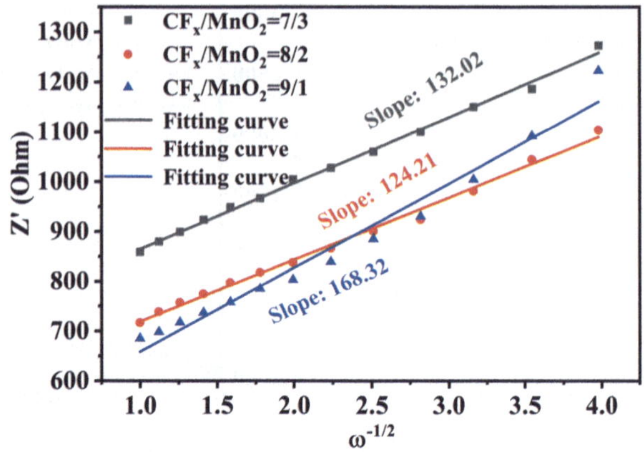

Fig. 8 Fitting relationship between Z' and $\omega^{-1/2}$

2.4 High- and Low-Temperature Performance Analysis of Fluorocarbon/Manganese Dioxide Mixed Positive Electrode

In order to test the high- and low-temperature performance of different proportions of CFx/MnO_2, the high- and low-temperature discharge test was carried out on the three kinds of proportion samples equipped with batteries. The high- and low-temperature discharge curves of electrode materials with CFx/MnO_2 ratio of 7/3, 8/2, and 9/1 at 0.1C are shown in Fig. 9.

When the temperature increases to 40 °C, the three samples all show a more obvious second discharge platform near 1.8 V, showing the over-theoretical capacities of 937.8, 1056.7, and 984 mAh·g^{-1}, which may be caused by the phase transition of active substances caused by the high-temperature discharge process. Among them, the sample with the proportion of 8/2 showed the specific capacity of 107.52, 240.64, 488.87, 909.85, and 1056.67 mAh·g^{-1} at −40 °C, −20 °C, 0 °C, 20 °C, and 40 °C, respectively, which was the best performance value among the three samples at various temperatures. Once again, it is confirmed that the CFx/MnO_2 ratio of 8/2 not only has excellent magnification performance but also has the best temperature adaptability. In addition, all three samples exhibit weak platform capacity and rapid voltage attenuation at temperatures below 0 °C, especially when the temperature drops to −40 °C, almost no platform capacity is displayed, and the main capacity is the slope capacity generated by surface adsorption. Therefore, it can be seen that the platform capacity of effective excitation should be controlled above −20 °C, and the higher the temperature between −20 and 40 °C, the higher the platform capacity of excitation.

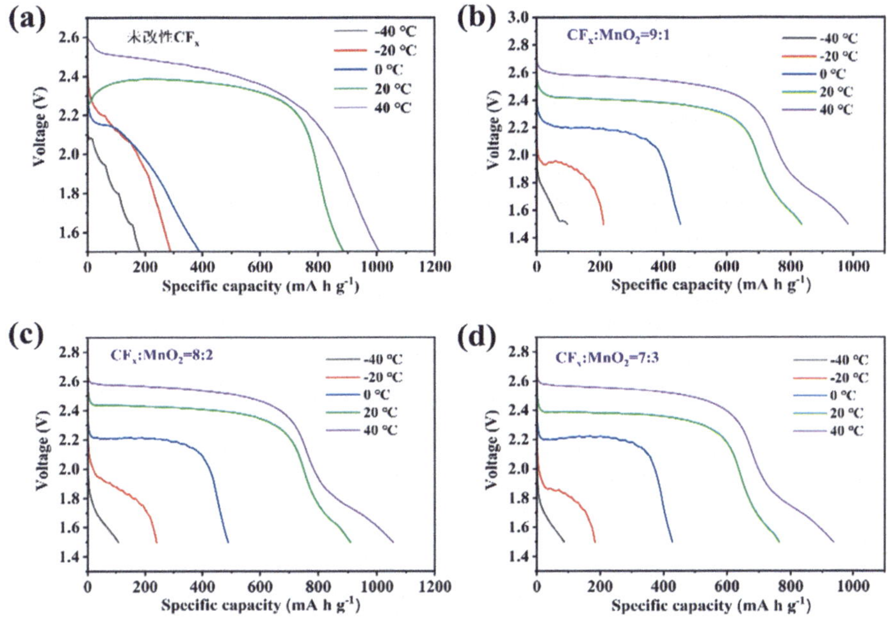

Fig. 9 Discharge graphs of CFx/MnO_2 with different mass ratios at different temperatures and rates of 0.1C: (**a**) Unmodified CFx, (**b**) 9/1, (**c**) 8/2, and (**d**) 7/3

The temperature-specific capacity diagram in Fig. 10 provides a detailed comparison of the temperature variation of discharge capacities of different proportions of MnO_2 and unmodified pure CFx samples. It can be seen that when the MnO_2 incorporation ratio is 8/2, the best discharge capacity is displayed in each temperature range. When the temperature is lower than −20 °C, the unmodified pure CFx shows little difference from the 8/2 ratio sample and is significantly higher than other ratio samples, indicating that the effect of MnO_2 incorporation is not obvious at too low temperature. When the temperature is 0 °C, the sample with MnO_2 ratio shows higher discharge capacity than the original sample. However, when the temperature is further increased to 20 °C, the specific discharge capacity of the original sample is further increased and exceeds that of the sample with the ratios of 9/1 and 7/3, indicating that the different incorporation amounts of MnO_2 have a great influence on the discharge capacity at normal temperature. Finally, tests at a temperature of 40 °C show that the introduction of a small amount of MnO_2 can significantly improve the high-temperature performance of CFx samples.

The temperature-median voltage diagram in Fig. 11 shows that, when the temperature is below 0 °C, the incorporation of a large proportion of MnO_2 will lead to the elevation of the discharge platform. When the temperature is higher than 0 °C, the platform voltage gap between the three will gradually decrease until it becomes the smallest at a high temperature of 40 °C.

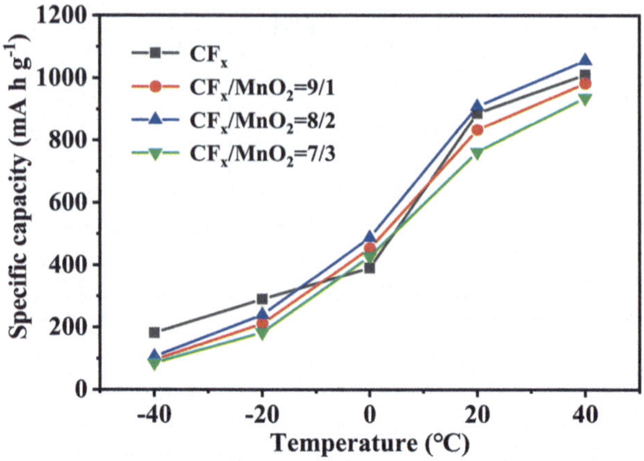

Fig. 10 Temperature-specific capacity diagram of unmodified CFx and CFx/MnO_2 with different mass ratios at different temperatures

Fig. 11 Temperature-median voltage diagram of CFx/MnO_2 with different mass ratios at 0.1C discharge rate

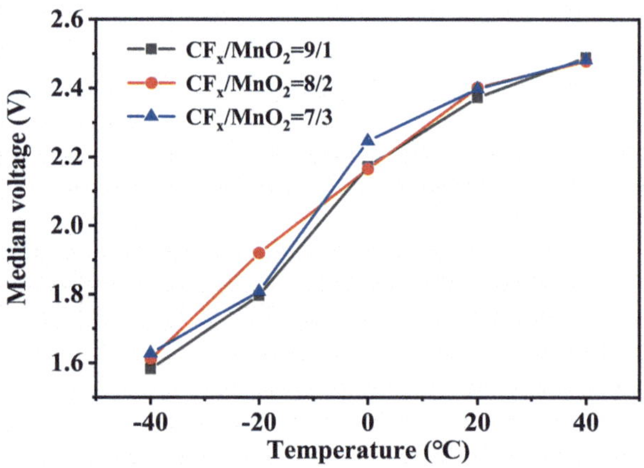

3 Summary

In this paper, CFx/MnO_2 composites with different MnO_2 contents were prepared by the solid-phase mixing method, and their phase structure and electrochemical properties were analyzed, and the following conclusions were drawn:

1. Tests at 20 °C with different ratios show that the discharge ratios of three kinds of CFx/MnO_2 composites with different mass ratios can all reach 5C, and the introduction of different ratios of MnO_2 has great differences in discharge performance. When the ratio of CFx to MnO_2 is 8/2, it has the best discharge performance at each rate.
2. When the ratio of CFx to MnO_2 is 8/2, the impedance value is significantly lower than that of the other two proportional samples, which indicates that

the introduction of an appropriate amount of MnO_2 can improve the overall electrical conductivity. The high- and low-temperature test at 0.1C shows that the contribution of MnO_2 to the discharge capacity is not obvious at low temperature. With the gradual increase of temperature, there is an obvious difference between the incorporated capacity and discharge platform of different proportions of MnO_2, in which the ratio of 8/2 has the best high- and low-temperature capacity.

References

1. Liu, Z., Su, H., Yang, Y., et al.: Advances and perspectives on transitional metal layered oxides for potassium-ion battery [J]. Energy Storage Mater. **34**, 211–228. Author, F., Author, S.: Title of a proceedings paper. In: Editor, F., Editor, S. (eds.) Conference 2016, LNCS, vol. 9999, pp. 1–13. Springer, Heidelberg (2016)
2. Zhang, C., Wang, F., Han, F., et al.: Improved electrochemical performance of sodium/potassium-ion batteries in ether-based electrolyte: cases study of MoS_2@C and Fe_7S_8@C anodes[J]. Adv. Mater. Interfaces **7**(13), 2000486 (2020). Author, F.: Contribution title. In: 9th International Proceedings on Proceedings, pp. 1–2. Publisher, Location (2010)
3. Xu, L., Xiong, P., Zeng, L., et al.: Facile fabrication of a vanadium nitride/carbon fiber composite for half/full sodium-ion and potassium-ion batteries with long-term cycling performance[J]. Nanoscale. **12**(19), 10693–10702 (2020)
4. Jiang, C., Li, X., Ying, Y., et al.: Fluorinated graphene-enabled durable triboelectric coating for water energy harvesting[J]. Small. **17**(8), e2007805 (2021)
5. Jones, J.-P., Jones, S.C., Billings, K.J., et al.: Radiation effects on lithium CFx batteries for future spacecraft and landers [J]. J. Power Sour. **471**, 228464 (2020)
6. Zhu, Y., Zhang, L., Zhao, H., et al.: Significantly improved electrochemical performance of CFx. Promoted by SiO2 modification for primary lithium batteries[J]. J. Mater. Chem. A. **5**(2), 796–803 (2017)

Effects of C-Rate on the Thermal Behavior of Commercial Lithium-Ion Pouch Cells with Different Aging Histories for a Successful Second-Life Transition

Emanuele Michelini, Patrick Höschele, Andreas Prielinger, Christian Ellersdorfer, and Jörg Moser

1 Introduction

As the number of batteries in automotive applications continues to rise [1], the need for effective strategies for managing these batteries at the end of their first life has become increasingly apparent. One promising solution is repurposing batteries for a second life [2]. Second-life batteries prolong the overall lifespan of the batteries [3], thereby reducing their environmental impact and increasing their economic efficiency [3, 4].

However, the transition from first life to second life introduces new challenges, with safety being a primary concern [5]. Altered operating conditions and usage patterns in the second application can lead to thermal, electrical, and mechanical loads that may not have been initially anticipated during their first-life battery design. In addition, batteries have already been subjected to many degradation mechanisms in their first life [6]. These degradation processes directly impact the battery behavior, causing safety-critical issues such as lithium plating [7] which coupled with the reduction in thermal conductivity [8] results in greater temperature inhomogeneities on the cell surface [9]. Furthermore, aged batteries are characterized by an increase in internal resistance [10], leading to increased heat generation due to ohmic effects [11]. Thus, it is opportune to test whether or not second-life loads are safety-critical.

One of the common stresses experienced by the battery is the electrical load, as the battery undergoes continuous charging and discharging with varying currents.

E. Michelini (✉) · P. Höschele · C. Ellersdorfer · J. Moser
Vehicle Safety Institute, Graz University of Technology, Graz, Austria
e-mail: e.michelini@tugraz.at

A. Prielinger
Fronius International GmbH, Wels-Thalheim, Austria

L. Chen (ed.), *Trends in Clean Energy Research*, Green Energy and Technology,
https://doi.org/10.1007/978-3-031-67987-2_18

The flow of current through the battery can lead to safety-critical effects, such as heat generation [6], swelling [12], and lithium plating [13]. These effects intensify as the current increases.

While previous studies have explored the impact of high currents on fresh cells [13–17], it is relevant to investigate this phenomenon in aged cells. Understanding how aging affects the safety criticality of electrical loads is essential for advancing our knowledge in this field and smoothening the transition from the first to the second life.

2 Method

The tested cells were 50 Ah NMC(111)/C pouch cells, with 3.0–4.2 V voltage range [18]. In the analysis, cells with diverse aging backgrounds were examined: F1, F2, A1, and A2, as shown in Table 1. Cells F1 and F2 were calendric aged and experienced minimal electrical cycling, whereas A1 and A2 underwent electrical cycling within commercial vehicles.

The batteries underwent a specific electrical load case, consisting of five charging cycles in constant current-constant voltage (CCCV) mode with C/20 cutoff current and five discharging cycles in constant current (CC) mode. The discharge and charge rates doubled each time after five cycles, starting at 0.5 C and ending at 4 C. This resulted in four electrical load cases, i.e., 0.5 C, 1.0 C, 2.0 C, and 4.0 C. K-thermocouples (class 1) placed in the center and near the positive and negative tabs continuously monitored the batteries' temperature throughout the entire testing process without the use of a thermal chamber or cooling system. During the electrical load cases to monitor the safety-criticality, a set of thermal parameters is examined. Specifically, the considered parameters include the following:

- The maximum absolute temperature measured by the thermocouples during a complete load case.
- The maximum temperature difference between the thermocouples positioned close to the tabs during a complete load case.
- The maximum temperature increase rate in °C/Ah at the end of each discharge during a complete load case.
- The median temperature measured in the center of the cell during a complete load case.

Table 1 Aging history of tested cells

Aging history	Calendar aging [years]	Cycling time [h]	State of Health
F1	7	0	99%
F2	10	8	99%
A1	8	1239	79%
A2	7	613	91%

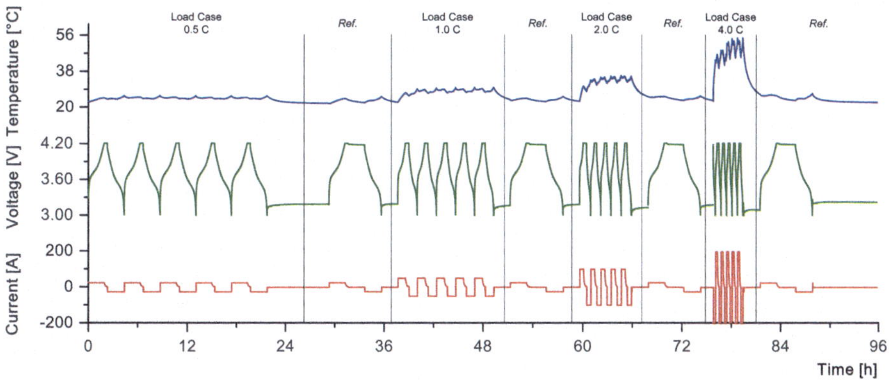

Fig. 1 Evolution of temperature, voltage, and current during the entire cycling protocol. The electric load cases are alternated with the reference basic charging-discharging profiles (*Ref.*)

- The interquartile range, namely, the difference between the temperature upper quartile and lower quartile, during a complete load case.

After each load case, the procedure presented by Michelini et al. [18] was used to extract a set of health indicators (HIs), namely, state of health (SOH), efficiency, voltage relaxation profile (VRP) voltage, VRP slope, area VdQ upon discharging, hysteresis open circuit voltage (OCV) approximation at 15 Ah (30% state of charge (SOC)) and 25 Ah (50% SOC), and hysteresis OCV approximation slope. The monitored HIs were used to assess the impact of the electrical load cases on the tested cells. The evolution of current, voltage, and temperature in the middle of the cell throughout the applied cycling protocol is shown in Fig. 1.

3 Results and Discussion

In Fig. 2 the temperature response corresponding to voltage changes during both charging and discharging processes is illustrated. The starting voltage at the beginning of charging differs because the cells discharged to 0% SOC with the same procedure relaxed to different voltages. Cycled aged cells A1 and A2 relax to a lower voltage.

The analysis of the thermal behavior indicates that higher absolute temperatures were observed during discharging. This phenomenon is attributed to increased internal resistance, leading to greater heat generation through the ohmic effect and larger kinetic limitations for lithiation compared to delithiation [19]. The cells, independently from aging, showed a repetitive thermal pattern upon charging and discharging. During charging, the temperature pattern can be divided into four key areas; see Fig. 2 (left). In Area 1 (from 3.00 to 3.60 V), the cell temperature exhibits a slight decrease. In Area 2 (from 3.60 to 3.85 V), occurring after the voltage knee,

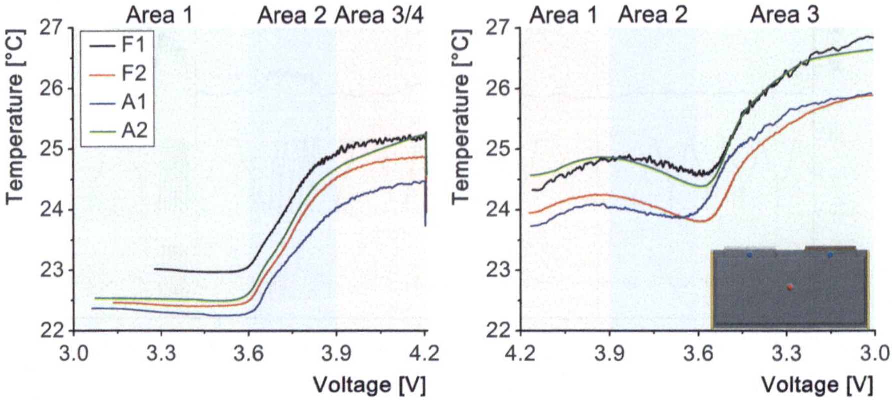

Fig. 2 Thermal evolution measured in the center of the cell over the voltage upon (left) charging and (right) discharging

exothermic reactions occur, causing a sharp increase in cell temperature. Area 3 (from 3.9 to 4.2 V) is characterized by a continued but less steady temperature rise. Finally, in Area 4 (at 4.2 V—CV phase), the temperature gradually decreases due to reduced heat generated from ohmic effects, as the current in the CV phase slowly decreases. Similarly, during discharging, the temperature pattern is categorized into three main areas; see Fig. 2 (right). Area 1 (from 4.2 to 3.9 V) maintains a nearly constant temperature, indicating stability. In Area 2 (from 3.9 to 3.7 V), the temperature slowly decreases, suggesting the potential presence of endothermic reactions absorbing more heat than the ohmic effect produces. In Area 3 (from 3.7 to 3.0 V), the temperature sharply increases, upon full cathode lithiation, thus SOC 0%. From the curves, it can be observed that the areas where the temperature increases during charging correspond with phases of constant or decreasing temperature during discharging and vice versa.

Analysis reveals that the majority of heat is generated at the end of the CC phase in both charging and discharging. This rise in heat production corresponds to increased electrode lithiation [20] and, especially upon discharging, to the increase in internal resistance [21]. However, the temperature increase upon charging is likely due to other factors, considering that the heat is primarily generated in the positive electrode and electrolyte [22]. Moreover, the internal resistance decreases at higher SOC, leading to a reduction in heat generated by ohmic effects [23]; thus, the temperature rise can be related to chemical exothermic reactions. The heat generated after reaching 3.7 V and then 3.9 V can be linked to electrochemical reactions following the voltage knee and the phase transition on the cathode side [24].

Table 2 summarizes the thermal parameters monitored during the load cases.

The thermal analysis has consistently revealed a proportional increase in the investigated thermal quantities in response to increasing C-rates, indicative of the

Table 2 Evolution of measured thermal parameters for the investigated cells

Thermal parameter	Cell	Electric load case			
		0.5 C	1.0 C	2.0 C	4.0 C
Max absolute temperature [°C]	F1	27.04	31.53	39.91	54.94
	F2	26.84	31.81	41.38	56.00
	A1	26.00	30.95	39.99	55.21
	A2	26.82	32.33	41.81	57.12
Max temperature difference [°C]	F1	1.64	2.77	4.44	7.48
	F2	1.10	1.96	3.38	7.95
	A1	0.76	1.59	3.10	6.95
	A2	1.24	2.40	4.14	8.19
Temperature increase rate [°C/ah]	F1	0.10	0.14	0.13	0.28
	F2	0.11	0.15	0.21	0.30
	A1	0.10	0.18	0.25	0.40
	A2	0.12	0.19	0.27	0.40
Median temperature [°C]	F1	25.03	28.60	35.38	48.71
	F2	24.85	28.95	37.06	49.24
	A1	24.25	28.12	35.41	48.15
	A2	24.76	29.02	36.79	49.33
Interquartile range [°C]	F1	0.60	1.38	2.54	5.40
	F2	0.65	1.47	3.48	6.84
	A1	0.67	1.52	3.69	7.39
	A2	0.59	1.52	3.69	7.39

direct influence of the current on the cell heat generation. This trend persists across all tested electrical load scenarios, irrespective of the SOH of the considered cell.

When examining cells with different aging histories, no definitive correlation emerges between the maximum temperature, maximum temperature difference, median temperature, and the aging process. While the electrically aged cell A2 often exhibits the highest heat generation, this pattern does not hold for the other electrically aged cell A1. In contrast, cell A1 consistently registers lower temperatures than its not electrically aged counterpart F2.

Two thermal parameters, namely, the temperature increase rate and the interquartile range, exhibit correlations with cell aging. The expansion of the interquartile range can be attributed to the diminishing thermal conductivity as the cell ages [8], making it more susceptible to temperature fluctuations due to the reduced thermal distribution. Additionally, the aging process reduces the cell's SOH [25], leading to a condensed period for exothermal reactions during the last discharge phase area (3.7–3.0 V). This condensed timeframe intensifies heat generation, increasing the temperature increase rate.

The effect of the electrical load cases was studied by monitoring the evolution of some HIs after applying the electrical load case, and the obtained results are summarized in Table 3.

Table 3 Evolution of the measured thermal parameters for the differently aged cells

Health indicator	Cell	Load case 0.5 C	1.0 C	2.0 C	4.0 C
State of health [%]	F1	99.11	99.23 (+0.12%)	99.28 (+0.06%)	98.45 (−0.84%)
	F2	99.32	99.36 (+0.04%)	99.41 (+0.05%)	98.09 (−1.33%)
	A1	79.26	79.55 (+0.37%)	79.54 (−0.01%)	77.87 (−2.11%)
	A2	91.00	93.13 (+2.35%)	92.77 (−0.40%)	88.39 (−4.71%)
Efficiency [%]	F1	99.09	99.07 (−0.02%)	98.77 (−0.30%)	97.97 (−0.81%)
	F2	99.12	99.01 (−0.11%)	98.66 (−0.35%)	97.79 (−0.88%)
	A1	96.99	98.33 (+1.37%)	98.20 (−0.13%)	97.84 (−0.37%)
	A2	97.40	98.73 (+1.36%)	98.69 (−0.03%)	97.87 (−0.84%)
Voltage relaxation profile [V]	F1	4.181	4.183 (+0.04%)	4.185 (+0.05%)	4.186 (+0.02%)
	F2	4.182	4.183 (+0.02%)	4.185 (+0.05%)	4.187 (+0.04%)
	A1	4.166	4.175 (+0.21%)	4.178 (+0.08%)	4.182 (+0.09%)
	A2	4.177	4.180 (+0.07%)	4.183 (+0.07%)	4.185 (+0.05%)
Voltage relaxation profile slope [μV/s]	F1	−0.85	−0.83 (−2.69%)	−0.67 (−19.35%)	−1.08 (+62.31%)
	F2	−0.82	−0.73 (−11.32%)	−0.78 (+6.25%)	−0.78 (+0.15%)
	A1	−1.64	−1.72 (+5.08%)	−1.26 (−26.54%)	−1.44 (+13.91%)
	A2	−0.97	−1.00 (+3.44%)	−0.89 (−11.40%)	−1.05 (+18.69%)
Area VdQ [VAh]	F1	185.05	185.49 (+0.24%)	185.87 (+0.20%)	184.61 (−0.68%)
	F2	185.51	185.77 (+0.14%)	186.12 (+0.19%)	184.03 (−1.13%)
	A1	152.82	153.59 (+0.50%)	153.83 (+0.16%)	151.06 (−1.80%)
	A2	174.94	174.11 (−0.47%)	173.67 (−0.25%)	170.9 (−1.59%)
OCV 25 Ah [V]	F1	3.688	3.685 (−0.09%)	3.682 (−0.07%)	3.683 (+0.03%)
	F2	3.686	3.684 (−0.07%)	3.681 (−0.06%)	3.685 (+0.11%)
	A1	3.781	3.784 (+0.09%)	3.782 (−0.07%)	3.798 (+0.43%)
	A2	3.704	3.708 (+0.11%)	3.706 (−0.05%)	3.716 (+0.28%)
OCV 15 Ah [V]	F1	3.603	3.599 (−0.12%)	3.595 (−0.11%)	3.595 (−0.01%)
	F2	3.601	3.598 (−0.09%)	3.594 (−0.10%)	3.596 (+0.05%)
	A1	3.643	3.644 (+0.05%)	3.640 (−0.11%)	3.648 (+0.21%)
	A2	3.608	3.61 (+0.04%)	3.606 (−0.12%)	3.611 (+0.14%)
Slope OCV 25–15 [Ah/V]	F1	118.04	116.97 (−0.91%)	115.17 (−1.54%)	113.16 (−1.74%)
	F2	117.37	116.16 (−1.03%)	114.5 (−1.43%)	111.62 (−2.51%)
	A1	72.54	71.6 (−1.29%)	70.86 (−1.04%)	66.78 (−5.75%)
	A2	104.70	101.99 (−2.58%)	99.8 (−2.15%)	94.58 (−5.24%)

From the evolution of the monitored His, some remarks can be made. Initially, at lower C-rates, there was no discernible negative impact on the SOH. However, following the 4 C load case, a decrease in SOH was evident across all cells, especially pronounced in the electrically aged cells A1 and A2. Concurrently, efficiency, lower in electrically aged cells, exhibited a consistent decrease after each load case, with the most substantial drop occurring after the highest load case, irrespective of the cell's aging history. In terms of the voltage relaxation profile, electrically aged cells consistently exhibited lower voltages, yet the load cases did not significantly alter the relaxed voltage. The voltage relaxation profile slope was influenced by cell aging, with electrically aged cells displaying a steeper voltage decrease after 1 h, particularly intensified after 4 C cycles. Furthermore, the VdQ area, an indicator directly associated with SOH as it reflects the capacity retained in the battery in Vah mirrored its trends, decreasing for all cells after the 4 C load case. Regarding the OCV at 25 and 15 Ah, values remained constant initially, but a slight increase was noted after the last load case, especially pronounced in electrically aged cells. Additionally, the slope between 25 and 15 Ah became progressively steeper after each applied load case, with higher load cases resulting in a more significant increase, particularly evident in electrically aged cells. The findings demonstrate a notable influence of the electrical load case on various HIs, with the 4 C profile being particularly impactful, especially for the electrically aged cells.

4 Conclusions

Given the criticality of cycling the cell with high currents, our study delved into studying the thermal behavior of cells with diverse aging histories cycled with different C-rates. The investigation revealed a consistent and repetitive thermal pattern exhibited by cells during charge and discharge cycles, a pattern intricately linked to the internal reactions occurring within the cell. The phase where the most significant heat is generated occurs toward the end of discharge, approximately between 3.6 and 3.0 V. This result is independent of the aging of the battery. Moreover, at higher C-rates, the heat generated by the cell increases significantly.

The study of thermal behavior as a function of C-rate and battery aging showed that while all thermal parameters are influenced by the applied C-rate, not all thermal parameters are directly influenced by aging. There appears to be no direct correlation between maximum absolute temperature, maximum temperature difference, median temperature, and battery aging. However, parameters such as the temperature increase rate and the interquartile range exhibited a clear correlation with the diminishing thermal conductivity and the declining SOH of the battery due to aging. As a validation, the impact of the electrical load case was assessed using HIs. The evolution of the HIs indicated that the load case with the highest C-rate had the most significant negative impact on the cell, and this was particularly pronounced in the electrically aged cells.

In conclusion, this study underscores the critical importance of considering the C-rate, particularly for aged cells. Aging cells exhibit an increased susceptibility to rapid and steep temperature rises, potentially leading to critical situations at an accelerated pace. Acknowledging and accounting for these thermal dynamics is paramount in ensuring the safe and effective utilization of batteries in real-world second-life applications.

Funding This work originates from the research project SafeLIB. The COMET Project SafeLIB is funded within the framework of COMET—Competence Centers for Excellent Technologies (Grant agreement No. 882506) by BMK, BMDW, the Province of Upper Austria, the province of Styria as well as SFG. The COMET Program is managed by FFG. The authors thank the consortium members of the SafeLIB project for supporting this work.

References

1. International Energy Agency (IEA): Global EV Outlook 2023: Catching up with climate ambitions. IEA (2023). [Online]. https://www.iea.org/reports/global-ev-outlook-2023
2. Automotive lithium-ion batteries considering technical, economic and legal aspects. Energies. **16**(6), 2830 (2023). https://doi.org/10.3390/en16062830
3. Mathews, I., Xu, B., He, W., Barreto, V., Buonassisi, T., Peters, I.M.: Technoeconomic model of second-life batteries for utility-scale solar considering calendar and cycle aging. Appl. Energy. **269**, 115127 (2020). https://doi.org/10.1016/j.apenergy.2020.115127
4. Kamath, D., Arsenault, R., Kim, H.C., Anctil, A.: Economic and environmental feasibility of second-life lithium-ion batteries as fast-charging energy storage. Environ. Sci. Technol. **54**(11), 6878–6887 (2020). https://doi.org/10.1021/acs.est.9b05883
5. Kampker, A., Heimes, H.H., Offermanns, C., Vienenkötter, J., Frank, M., Holz, D.: Identification of challenges for second-life battery systems – a literature review. WEVJ. **14**(4), 80 (2023). https://doi.org/10.3390/wevj14040080
6. Barré, A., Deguilhem, B., Grolleau, S., Gérard, M., Suard, F., Riu, D.: A review on lithium-ion battery ageing mechanisms and estimations for automotive applications. J. Power Sources. **241**(9), 680–689 (2013). https://doi.org/10.1016/j.jpowsour.2013.05.040
7. Liu, J., Zhou, L., Zhang, Y., Wang, J., Wang, Z.: Aging behavior and mechanisms of lithium-ion battery under multi-aging path. J. Clean. Prod. **423**, 138678 (2023). https://doi.org/10.1016/j.jclepro.2023.138678
8. Kovachev, G., et al.: Thermal conductivity in aged Li-ion cells under various compression conditions and state-of-charge. Batteries. **7**(3), 42 (2021). https://doi.org/10.3390/batteries7030042
9. Xie, Y., et al.: Inhomogeneous degradation induced by lithium plating in a large-format lithium-ion battery. J. Power Sources. **542**, 231753 (2022). https://doi.org/10.1016/j.jpowsour.2022.231753
10. Barcellona, S., Colnago, S., Dotelli, G., Latorrata, S., Piegari, L.: Aging effect on the variation of Li-ion battery resistance as function of temperature and state of charge. J. Energy Storage. **50**, 104658 (2022). https://doi.org/10.1016/j.est.2022.104658
11. Galatro, D., Al-Zareer, M., Da Silva, C., Romero, D.A., Amon, C.H.: Thermal behavior of lithium-ion batteries: aging, heat generation, thermal management and failure. Front. Heat Mass Transf. **14** (2020). https://doi.org/10.5098/hmt.14.17
12. Bitzer, B., Gruhle, A.: A new method for detecting lithium plating by measuring the cell thickness. J. Power Sources. **262**(36), 297–302 (2014). https://doi.org/10.1016/j.jpowsour.2014.03.142
13. Münster, P., et al.: Effect of Li plating during formation of lithium ion batteries on their cycling performance and thermal safety. J. Power Sources. **484**, 229306 (2021). https://doi.org/10.1016/j.jpowsour.2020.229306
14. Dong, T., Peng, P., Jiang, F.: Numerical modeling and analysis of the thermal behavior of NCM lithium-ion batteries subjected to very high C-rate discharge/charge operations. Int. J. Heat Mass Transf. **117**, 261–272 (2018). https://doi.org/10.1016/j.ijheatmasstransfer.2017.10.024
15. Zhang, G., Cao, L., Ge, S., Wang, C.-Y., Shaffer, C.E., Rahn, C.D.: In situ measurement of radial temperature distributions in cylindrical Li-ion cells. J. Electrochem. Soc. **161**(10), A1499–A1507 (2014). https://doi.org/10.1149/2.0051410jes

16. Gao, Z., Xie, H., Yang, X., Niu, W., Li, S., Chen, S.: The dilemma of C-rate and cycle life for lithium-ion batteries under low temperature fast charging. Batteries. **8**(11), 234 (2022). https://doi.org/10.3390/batteries8110234
17. Wang, L., et al.: Construction of electrochemical model for high C-rate conditions in lithium-ion battery based on experimental analogy method. Energy. **279**, 128073 (2023). https://doi.org/10.1016/j.energy.2023.128073
18. Michelini, E., Höschele, P., Abbas, S.M., Ellersdorfer, C., Moser, J.: Assessment of health indicators to detect the aging state of commercial second-life lithium-ion battery cells through basic electrochemical cycling. Batteries. **9**(11), 542 (2023). https://doi.org/10.3390/batteries9110542
19. Manikandan, B., Yap, C., Balaya, P.: Towards understanding heat generation characteristics of Li-ion batteries by calorimetry, impedance, and potentiometry studies. J. Electrochem. Soc. **164**(12), A2794–A2800 (2017). https://doi.org/10.1149/2.1811712jes
20. Maleki, H., Deng, G., Kerzhner-Haller, I., Anani, A., Howard, J.N.: Thermal stability studies of binder materials in anodes for lithium-ion batteries. J. Electrochem. Soc. **147**(12), 4470 (2000). https://doi.org/10.1149/1.1394088
21. Liu, G., Ouyang, M., Lu, L., Li, J., Han, X.: Analysis of the heat generation of lithium-ion battery during charging and discharging considering different influencing factors. J. Therm. Anal. Calorim. **116**(2), 1001–1010 (2014). https://doi.org/10.1007/s10973-013-3599-9
22. Du, S., et al.: An investigation of irreversible heat generation in lithium ion batteries based on a thermo-electrochemical coupling method. Appl. Therm. Eng. **121**, 501–510 (2017). https://doi.org/10.1016/j.applthermaleng.2017.04.077
23. Lyu, P., Huo, Y., Qu, Z., Rao, Z.: Investigation on the thermal behavior of Ni-rich NMC lithium ion battery for energy storage. Appl. Therm. Eng. **166**, 114749 (2020). https://doi.org/10.1016/j.applthermaleng.2019.114749
24. Fröhlich, K., Abrahams, I., Jahn, M.: Determining phase transitions of layered oxides via electrochemical and crystallographic analysis. Sci. Technol. Adv. Mater. **21**(1), 653–660 (2020). https://doi.org/10.1080/14686996.2020.1814116
25. Tian, H., Qin, P., Li, K., Zhao, Z.: A review of the state of health for lithium-ion batteries: research status and suggestions. J. Clean. Prod. **261**, 120813 (2020). https://doi.org/10.1016/j.jclepro.2020.120813

Research on Preparation Process of Graphene Oxide of Negative Electrode Materials

Deng Hui-Yong, Sun Hai-Tao, Cui Kai-Bo, and Liu-Liang

1 Introduction

In recent years, the development of the new energy vehicle industry and the huge demand in the energy storage field have driven the rapid development of battery technology materials. This presents a huge opportunity and challenge for the improvement of battery material preparation technology and the study of energy storage mechanisms. From the development of batteries, their electrochemical performance mainly depends on the structure and performance of the electrode and electrolyte materials used, especially the selection of positive and negative electrode materials. The safety of negative electrode materials directly determines the promotion and application of battery products. Therefore, the development of inexpensive and high-performance negative electrode materials has always been a focus of battery research [1].

Graphite is currently the most commercialized negative electrode material for lithium-ion batteries [2]. Graphite has a layered structure, with carbon atoms existing in the form of sp^2 or sp^3 hybridization, forming a layered graphene plane with carbon-carbon double bonds. In graphite, carbon atoms in the same layer are arranged in a regular hexagonal shape, and the layers are bonded by van der Waals forces [3, 4]. Graphite has good conductivity, high crystallinity, a stable charging and discharging platform, and advantages such as long cycle time, low cost, and abundant resources. However, both its specific gravity capacity and specific volume capacity are low. According to the difference in the degree of graphitization of materials, mature carbon negative electrode materials can usually be divided into soft carbon, hard carbon, and graphite. Traditional carbon materials exhibit issues such as low reversible capacity, poor rate performance, and incompatibility with

D. Hui-Yong · S. Hai-Tao · C. Kai-Bo · Liu-Liang (✉)
Army Engineering College of PLA, Hebei, China

L. Chen (ed.), *Trends in Clean Energy Research*, Green Energy and Technology,
https://doi.org/10.1007/978-3-031-67987-2_19

electrolytes, leading to decreased cycling performance. Graphene oxide, such as graphene, is a two-dimensional nanomaterial. Oxidized graphene is essentially a honeycomb such as single atomic layer carbon atomic crystal composed of carbon atoms in the form of sp^2 hybridization. The surface and edges of its single layer carbon atomic layer are connected with many oxygen-containing functional groups, including hydroxyl, epoxy, and carboxyl groups. The commonly used method for preparing graphene oxide is the improved Hummers method [5], which involves oxidizing graphite powder with concentrated sulfuric acid and potassium permanganate to obtain graphene oxide and then subjecting it to water or other suitable solvents for ultrasonic exfoliation to obtain graphene oxide. In addition, graphene can also be obtained by first treating expanded graphite or ball milling and then oxidizing graphene to obtain oxidized graphene [6]. In this paper, $KMnO_4$ was used instead of $NaNO_3$ to prepare oxidized graphene using the Hummers method, and the performance of $KMnO_4$ in preparing oxidized graphene was tested through experimental means.

2 Preparation of Graphene Oxide

2.1 Preparation of Graphene Oxide

Prepare graphene oxide without adding $NaNO_3$ according to the steps of the Hummers method. The process is as follows: slowly pour 70 mL of concentrated H_2SO_4 into a 0.5 L beaker, vigorously stir with a magnet, and cool in an ice water bath. Then add powdered natural graphite flakes (3.0 g) to the stirred and cooled sulfuric acid. After stirring until the graphite flakes are evenly dispersed in the system, slowly shake $KMnO_4$ (9.0 g) with a spoon and add it to the beaker. Be careful not to directly add $KMnO_4$ to the vortex center of the magnetic stirring, nor stick it to the beaker wall. This process takes about 20 min and the temperature does not exceed 20 °C. After adding, stir evenly until no solid $KMnO_4$ particles are visible. Then transfer the reaction system to an oil bath that has been kept at a constant temperature of 40 °C, cover it with glass sheets, and vigorously stir for 30 min. After the intermediate temperature oxidation process at 40 °C is completed, slowly add deionized water (150 mL) to the oxidation system to avoid excessive heat release and rapid gas generation in the reaction system; otherwise, it may cause the reaction solution to overflow the beaker. After adding water, heat up to 95 °C and continue to react at this temperature for 15 min. After the oxidation process is completed, slowly pour the reaction solution into a 1000 mL beaker containing 500 mL of deionized water, and then add hydrogen peroxide (15 mL, 30%) dropwise. At this point, the color of the suspension changes from brown to yellow. After filtering the product with two layers of medium-speed qualitative filter paper, the filter cake is washed with hydrochloric acid in a 1:10 diluted aqueous solution (250 mL) to remove residual Mn and K ions. After drying the filter cake in the air, peel it off, tear it apart, and stir it overnight with 600 mL of deionized water to obtain the dispersion of the oxidized stone ink system. Dialyze this dispersion in a 27-mm-wide dialysis bag with a molecular weight of 8000–14,000 for at least 7 days, and change the dialysis water daily. Dilute the dialyzed graphene oxide solution to 1200 mL, stir overnight, and sonicate for 30 min to obtain a dispersed graphene oxide solution. Finally, use a centrifuge with a speed of 3000 r/min to remove the untreated and poorly oxidized graphite particles.

2.2 *Reduction of Graphene Oxide*

Dry approximately 5 mg/mL of GO solution naturally to obtain a GO film. Then, the reduced graphene oxide (rGO) film was obtained by overnight soaking with hydroiodic acid, and finally washed with water and ethanol until no yellow iodine was dissolved and further dried.

3 Structural Characterization Methods

Raman spectroscopy was collected under experimental conditions of 4.7 mW with a 514 nm laser on a Renishaw Raman spectrometer. X-ray photoelectron spectroscopy (XPS) was collected using the ESCALAB 250 photoelectron spectrometer produced in the United States. Using 150 W strength Al for K α, the (1486.6 eV) spectral line is used as an X-ray source for high-resolution scanning with an energy of 30 eV. X-ray diffraction spectroscopy (XRD) is performed on an X-ray diffractometer using Cu K a line (wavelength $\lambda = 1.5418$ A) radiation acquisition. The UV-visible spectrum was collected on a U-3010 UV-visible spectrometer. The scanning electron microscopy (SEM) image is taken by placing GO low concentration dispersed droplets on a silicon wafer and using Sirion-200 scanning electron microscopy. The atomic force microscopy (AFM) image was obtained on the Shimadzu SPM-9600 scanning probe microscope. The sample preparation method was to drop low concentration dispersed liquid onto newly split mica sheets and air-dry them naturally. Attenuated total reflection Fourier transform infrared spectroscopy (ATR-FTIR) was measured on the Vertex V70 Fourier transform infrared spectrometer from Bruker Brooke, Germany. The zeta potential was measured on the HORIBA SZ-100 nanoparticle analyzer.

4 Performance Analysis of Preparing Graphene Oxide

For the convenience of discussion, we will analyze graphene oxide prepared by the traditional $NaNO_3$ method and $KMnO_4$ method. The dry GO used for structural characterization is obtained by freeze-drying the corresponding GO solution for 48 h.

4.1 Spectral Analysis of Graphene Oxide

The GO yield prepared by the Hummers method using $KMnO_4$ (the mass of GO divided by the mass of graphite raw material) is 92.3%. The corresponding traditional Hummers method resulted in a GO yield of 96 ± 2%. The similar yield proves that the reaction solution composed of sulfuric acid and potassium permanganate has the ability to achieve similar oxidation effects as the reaction solution composed of sulfuric acid, sodium nitrate, and potassium permanganate and can also be used to prepare GO.

After confirming similar GO yields, detailed structural characterization was performed on the GO prepared by the two Hummers methods. The 232 nm peak and its 300 nm shoulder peak in the UV-visible spectrum of the GO aqueous solution of $KMnO_4$ in Fig. 1a correspond to the π-π^* transition of the C=C double bond and the n-π^* transition of the C=O double bond, respectively. The UV spectra of GO prepared by the $KMnO_4$ method are almost identical to those of GO prepared by $NaNO_3$ (Fig. 2a), both of which are typical UV spectra of GO reported in the literature. The GO dispersions in Figs.1a and 2a are both clear yellow, confirming that they are products of fully oxidizing and functionalizing graphite. The C/O atomic ratio calculated in the XPS spectrum is also very close, with GO values of 2.36 and 2.23 for $KMnO_4$ and traditional GO with $NaNO_3$, respectively (Figs. 1b and 2b). This C/O atomic ratio falls within the atomic ratio range of successfully prepared GO. The C1s spectra of XPS (Figs. 2c and 3c) show four types of carbon atoms, namely, C-C/C=C(284.8 eV), C-O-C/C-OH(286.8 eV), C=O(287.8 eV), and O-C=O (289.0 eV). The proportion of complete carbon (C-

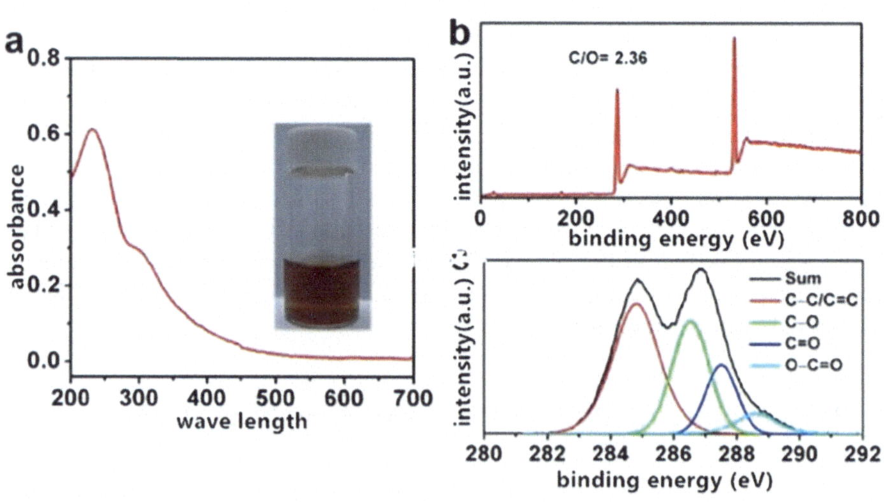

Fig. 1 (**a**) UV-visible spectrum, (**b**) XPS full-spectrum scan, (**c**) C 1 s scan spectrum

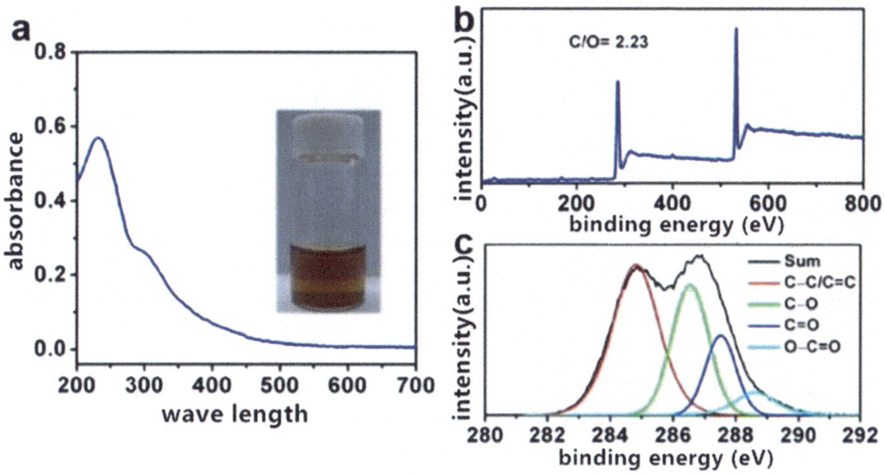

Fig. 2 (**a**) UV-visible spectrum, (**b**) XPS full-spectrum scan, (**c**) C 1 s scan spectrum

C/C=C) and oxidized carbon (C-O-C/C-OH, C=O and O-C=O) calculated based on the peak division of the GO spectrum of $KMnO_4$ is 47.9 and 52.1%, respectively. The GO values of the traditional $NaNO_3$ are 46.5 and 53.5%, respectively. Although the oxidation degree of GO prepared by the traditional $NaNO_3$ method is slightly higher than that prepared by the $KMnO_4$ method, the oxidation degree of both methods is at the intermediate level of reported values.

In addition, the zeta potentials of GO containing $KMnO_4$ and GO containing $NaNO_3$ in the dispersion were -43.8 ± 1.3 and -45.6 ± 0.6 mV, respectively. This confirms that both have a significant negative charge and are the result of ionization of carboxyl and sulfate groups. Although the GO of $KMnO_4$ has relatively less negative charge, the zeta potential values of both are lower than the critical value of -30 mV required for stable dispersion in water, indicating that both $KMnO_4$ GO and the traditional GO can be stably dispersed in water.

4.2 *Structural Analysis of Graphene Oxide*

As shown in Figs. 3a and 4a, the thickness of GO without $NaNO_3$ and GO with $NaNO_3$ was measured by AFM, both around 0.8 nm, confirming that they are all single-layer structures. The GO layers produced by the two methods have similar sizes, ranging from a few hundred nm to a few hundred nm and μm, as shown in SEM photos in Figs. 3b and 4b. The XRD diffraction spectrum (as shown in Fig. 3c) shows that GO without NaNO3 exhibits a diffraction angle of 2 θ=. The diffraction peak at 10.9° corresponds to an interlayer spacing of 0.81 nm, which is basically consistent with the results obtained from GO with $NaNO_3$ (Fig. 4c). The

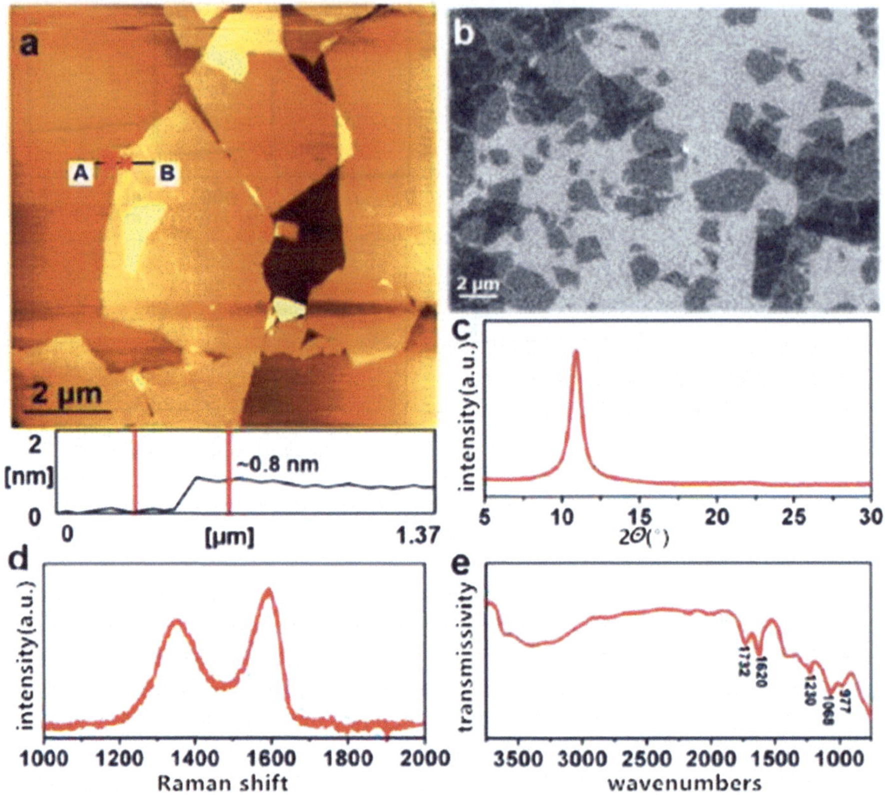

Fig. 3 (**a**) AFM scanning photos, (**b**) SEM photos, (**c**) XRD diffraction spectra, (**d**) Raman spectra, (**e**) total reflectance infrared spectra

larger interlayer spacing of GO prepared by the two methods is due to the oxygen-containing functional groups introduced by the strong oxidation reaction, which widens the interlayer spacing of the original graphite.

Raman spectroscopy and infrared spectroscopy also confirmed the structural similarity of GO prepared by the two methods. Both the Raman spectra of GO without $NaNO_3$ (Fig. 3d) and GO with $NaNO_3$ (Fig. 3d) show G peaks at approximately 1590 cm^{-1} and D peaks at approximately 1350 cm^{-1}. The G peak corresponds to graphitized carbon, while the D peak corresponds to structural defects or partially random graphite regions. The D peak of both is strong and the G peak is wide, indicating that their graphite structures have been significantly damaged due to functionalization. On the other hand, infrared spectroscopy confirmed the presence of many oxygen-containing functional groups in GO without $NaNO_3$ (Fig. 3e) and GO with $NaNO^3$ (Fig. 4e), for example, C-O-C (1000 cm^{-1}), C-O (1230 cm^{-1}), O-H bending vibration of bound water (1620 cm^{-1}), C=O (1740–1720 cm^{-1}),

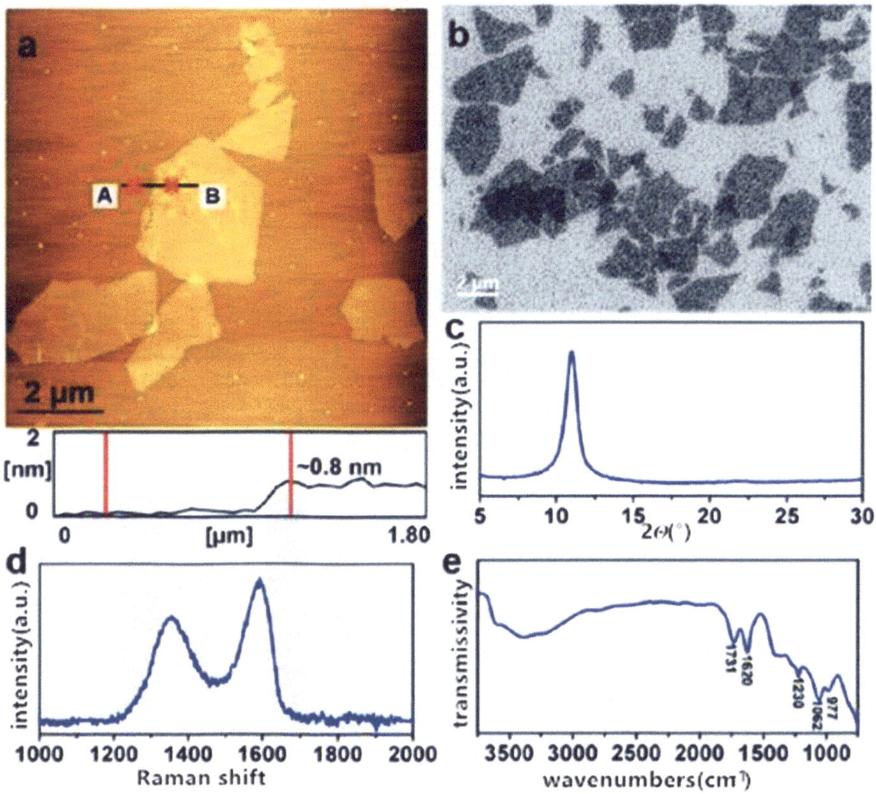

Fig. 4 (**a**) AFM scanning photos, (**b**) SEM photos, (**c**) XRD diffraction spectra, (**d**) Raman spectra, (**e**) total reflectance infrared spectra

and carboxyl, hydroxyl, and interlayer bound water and O-H stretching vibration (3600–3300 cm^{-1}). These hydrophilic oxygen-containing functional groups ensure the good dispersibility of GO in water.

4.3 *Thermogravimetric Analysis of Graphene Oxide*

Thermogravimetric analysis (TGA) can determine the amount of oxygen-containing functional groups in GO and their decomposition temperature. As shown in Fig. 5, the TGA curves measured by GO prepared with or without sodium nitrate method highly overlap, indirectly confirming their structural similarity. The specific analysis is as follows: the weight loss before 100 °C mainly corresponds to the escape of water confined between the layers of GO sheets. The significant weight loss at 200–230 °C corresponds to the decomposition of unstable oxygen-containing functional

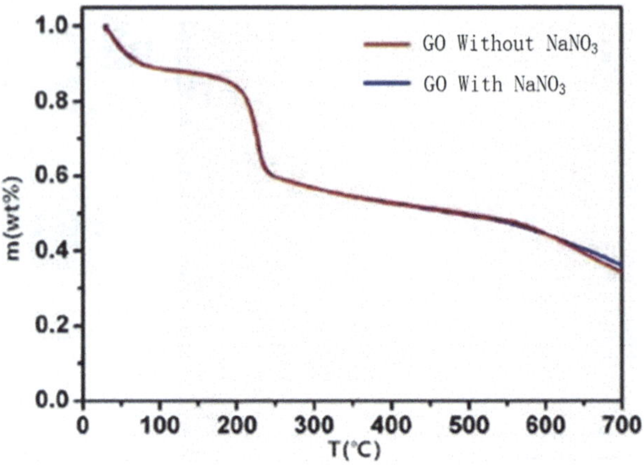

Fig. 5 Thermogravimetric analysis of without $NaNO_3$ and with $NaNO_3$

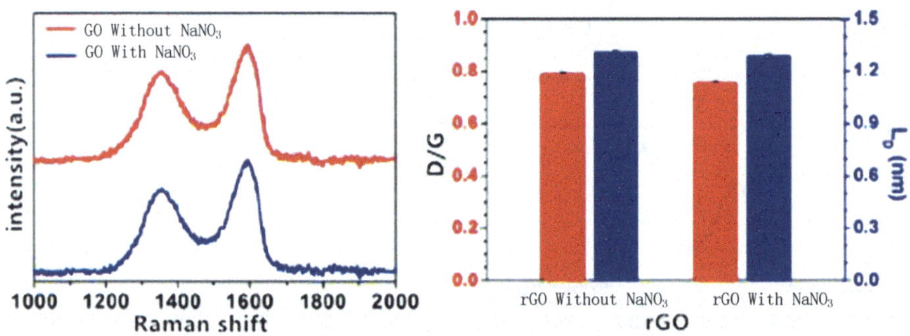

Fig. 6 Raman spectrum and comparison of D/G intensity ratio and LD value of GO with and without $NaNO_3$

groups on the GO layer. The slower weight loss at 300–600 °C corresponds to the relatively more stable decomposition of oxygen-containing functional groups.

4.4 Structural Integrity Analysis of Graphene Oxide

As shown in Fig. 6, the D/G intensity ratio in the Raman spectrum of GO without $NaNO_3$ is 0.79, while the corresponding value for GO with $NaNO_3$ is 0.76. Due to the fact that GO and rGO materials often belong to the II region of graphene materials, the higher the D/G ratio, the larger the average defect distance LD, and the lower the defect density. We calculated that the LD values of GO without $NaNO_3$ and GO with $NaNO_3$ were 1.30 nm and 1.26 nm, respectively, based on the relationship between D/G intensity ratio and LD. But here we cannot conclude

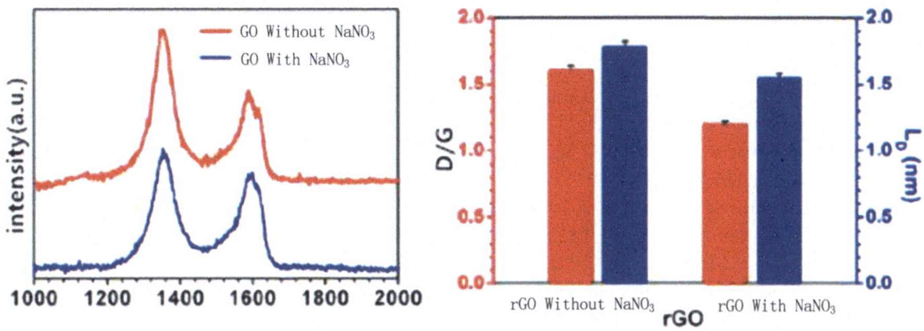

Fig. 7 Raman spectrum and comparison of D/G intensity ratio and LD value of rGO with $NaNO_3$

that GO without $NaNO_3$ has better structural integrity, because the defects reflected by LD can be two types: so-called defects caused by recoverable functional groups and permanent and irreversible defects. Therefore, we need to compare the LD values after reducing and removing functional groups to determine the number of permanent defects.

After GO is reduced by hydroiodic acid, most of the oxygen-containing functional groups on it are effectively removed, and the recoverable graphite structure is basically restored. At this point, comparing LD values can accurately determine the number of permanent defects. As shown in Fig. 7, the Raman spectra of rGO without $NaNO_3$ and rGO with $NaNO_3$ show significant differences. The D peak intensity of rGO without NaNO3 is significantly stronger than the G peak intensity, with a D/G intensity ratio of 1.6, corresponding to a LD value of 1.78 nm. The D/G intensity and LD value ratio of rGO with $NaNO_3$ are only 1.2 and 1.55 nm. At this point, we can determine that the GO and its reduction product rGO prepared by the method without $NaNO_3$ have higher structural integrity and fewer permanent defects compared to the GO and its reduction product prepared by the method with $NaNO_3$.

5 Conclusion

Using KMnO4 instead of $NaNO_3$ to prepare GO by the Hummers method has good consistency compared to traditional methods in terms of oxidation degree, stable dispersion in water, and thermogravimetric analysis. However, the GO of $KMnO_4$ has a D/G strength of 1.2 and an LD value of 1.55 nm, which is 1.6 and 1.78 nm compared to the GO prepared by traditional methods. This indicates that the GO structure of $KMnO_4$ has higher integrity and fewer permanent defects, which has great significance for preparing higher-quality GO.

References

1. Hong, C., Yu-qing, C.: Research status of negative electrode materials for lithium ion batteries. Chem. Energy. **49**(4), 148–150 (2023)
2. Reddy, M.V., Mauger, A., Julien, C.M., et al.: Brief history of early lithium-battery development. Materials. **13**(8), 1884 (2020)
3. Chung, D.D.L.: Review graphite. J. Mater. Sci. **37**(8), 1475–1489 (2002)
4. Song, X.Y., Kinoshita, K., Tran, T.D.: Microstructural characterization of lithiated graphite. J. Electrochem. Soc. **143**(6), L120 (1996)
5. Mitra, S., Banerjee, S., Datta, A., et al.: A brief review on graphene/inorganic nanostructure composites: materials for the future. Ind. J. Phys. **90**(9), 1019–1032 (2016)
6. Wang, Y., Tian, M., Qu, L.: Research on the status and development trend of graphene. J. Chengdu Textile College. **33**(1), 1–18 (2016)

Energy Storage Capacity of Microencapsulated Phase Change Materials

Rubén D. Santiago-Acosta, José A. Otero, and Ernesto M. Hernández-Cooper

1 Introduction

The problem of greenhouse gas emissions has become a major concern due to its contribution to global warming. One of the many alternatives to mitigate these emissions that result from fossil fuel consumption consists of harvesting solar radiation for thermoelectric generation in thermo-solar plants [1]. The challenges related to satisfy the energy-supply demand chain in these types of applications consist of materials' research to increase the energy density of backup systems and development of devices to enhance heat transfer rates [2–6]. Solar radiation is a highly intermittent source of energy since it is related to weather conditions and seasonal variations. Generally speaking, a short time window of direct solar irradiance is available for thermal energy storage. Phase change processes constitute a way to increase energy densities through the latent heat stored in first-order phase transitions. Materials that experience this type of transition during thermal energy storage are known as phase change materials (PCMs). Typically, these materials have very low thermal conductivities, which constitutes a major disadvantage to meet the required energy transfer rates in backup systems. The use of fins as heat exchangers [2], flip and rotating mechanisms to avoid uneven melting [4, 6], PCM composites [5], and microencapsulation of PCMs [7] to increase the heat transfer surface has been developed to reduce melting times.

Microencapsulated materials for solar energy harvesting at high temperatures consist of a PCM core, typically a salt, and a metallic shell with high thermal conductivity for rapid heat transfer rates [8–10]. The physics of melting processes in confined systems, such as spherically encapsulated PCMs, has been addressed

R. D. Santiago-Acosta · J. A. Otero · E. M. Hernández-Cooper (✉)
Tecnologico de Monterrey, Escuela de Ingeniería y Ciencias, Monterrey, Mexico
e-mail: emcooper@tec.mx

L. Chen (ed.), *Trends in Clean Energy Research*, Green Energy and Technology,
https://doi.org/10.1007/978-3-031-67987-2_20

by several authors [9–12]. Some of these authors have developed simple heat conduction models, but avoid changes in the liquid-solid saturation line by considering phases with equal densities [12]. Other authors have described the phase change by assuming incompressible phases [8–10]. The stress at the core-shell interface can be greatly overestimated through the application of models close to the incompressible limit, and melting times and energy densities can be also estimated incorrectly in this limit [13]. In this work, the energy density and melting times of compressible salts are estimated in two types of systems: free shell's outer surface and microencapsulated PCM composite-like system with a fixed shell-matrix interface. In contrast to what might be expected in latent heat storage physics, where encapsulation tends to decrease latent heat, it is found that energy density is greatly enhanced through encapsulation of salts with high compressibilities. Additionally, depending on the elastic and thermodynamic properties of the PCM, energy densities are enhanced in PCM composite such as encapsulates. Finally, melting times tend to increase with the shell's thickness, and these are found to be particularly higher in composite-like systems.

2 Physical Description and Main Assumptions

Microencapsulated PCMs are constituted by a PCM core of radius r_0 and a spherical shell of initial thickness $\delta_0 = R_0 - r_0$, where R_0 is the radius of the outer surface of the shell. Initially, the core is mostly in its solid state, where the initial radius of the solid sphere is r_{s_0}. Two types of microencapsulates are considered in this work: a shell with a free-moving outer surface that resembles encapsulates immersed in a medium at ambient pressure and a shell with a fixed outer surface that resembles microcapsules embedded in a matrix. The following assumptions are made for model development: (a) Young's modulus of the matrix is much higher than Young's modulus of the shell's material which enables the application of a fixed outer surface boundary condition; (b) initially, the PCM is mostly in its solid state with an initial arbitrary temperature profile that meets the isothermal boundary condition at the solid-liquid interface; (c) thermodynamic equilibrium is assumed to estimate the liquid-solid saturation line (overheating of solid is not considered); (d) system size is small enough, $r_0 \leq 1$ mm, so that heat transfer within the liquid phase lies in the conductive regime; (e) compressibility, specific heat, and thermal conductivity of the liquid (solid) phase are independent of temperature and pressure; (f) thermal expansion effects are not considered; and (g) the melting process is radially symmetric.

3 Shell's Elastic Properties

The elastic properties of the shell's material will be briefly described in this section. The shell's material is assumed to be isotropic and homogeneous. The stress, strain, and displacement relations in spherical coordinates are radially symmetric [8].

The assumed symmetry leads to the following differential equation for the radial displacement u_r within the shell's material:

$$\frac{d^2u_r}{dr^2} + \frac{2}{r}\frac{du_r}{dr} - \frac{2}{r^2}u_r = 0. \tag{1}$$

The boundary conditions in terms of Lamé's constants λ and μ for a free outer surface are given as follows:

$$(\lambda + 2\mu)\left.\frac{du_r}{dr}\right|_{r=r_0} + 2\lambda\frac{u_r(r_0)}{r_0} = -\Delta P, \tag{2}$$

$$(\lambda + 2\mu)\left.\frac{du_r}{dr}\right|_{r=R_0} + 2\lambda\frac{u_r(R_0)}{R_0} = 0. \tag{3}$$

Here, ΔP represents the relative pressure in the PCM. Alternatively, the boundary conditions for a microencapsulated salt embedded in a matrix are given as follows:

$$(\lambda + 2\mu)\left.\frac{du_r}{dr}\right|_{r=r_0} + 2\lambda\frac{u_r(r_0)}{r_0} = -\Delta P,$$

$$u_r(r_0) \tag{4}$$

The outer surface of the shell is a fixed boundary and represents the matrix-shell interface. The differential equation for the radial displacements has an exact analytical solution, which gives the following result for the confined PCM in a spherical shell with the boundary conditions shown through Eqs. (2)–(3):

$$u_r(r) = \frac{r_0^3}{r^2}\frac{\Delta P}{4\mu}\frac{\left((3\lambda + 2\mu) + 4\mu r^3/R_0^3\right)}{(3\lambda + 2\mu)\left(1 - r_0^3/R_0^3\right)}. \tag{5}$$

The displacements for a PCM embedded in a matrix can be obtained by applying the boundary conditions shown in Eq. (4), to the analytical solution of Eq. (1) as follows:

$$u_r(r) = \frac{r_0^3}{r^2}\frac{\Delta P\left(1 - r^3/R_0^3\right)}{4\mu + (3\lambda + 2\mu)\, r_0^3/R_0^3}. \tag{6}$$

The displacement of the inner surface in each case can be obtained through evaluation of Eqs. (5) and (6) at $r = r_0$, which leads to the following expressions:

$$u_r(r_0) = r_0 a_1 p_d \text{ with } a_1 = \frac{1 + \nu + (2 - 4\nu)\, r_0^3/R_0^3}{2\left(1 - r_0^3/R_0^3\right)} \tag{7}$$

$$u_r(r_0) = r_0 a_2 p_d \text{ with } a_2 = \frac{(1+\nu)(1-2\nu)\left(1 - r_0^3/R_0^3\right)}{(2-4\nu) + (1+\nu)\, r_0^3/R_0^3}. \tag{8}$$

Here, p_d is a dimensionless relative pressure, and it is expressed in multiples of Young's modulus E of the shell's material as $p_d = \Delta P/E$. The definition of Lamé's constants in terms of Young's modulus E and Poisson's ratio ν of the shell was used to find the displacement of the PCM-shell interface. Equations (7) and (8) show the displacement of the PCM-shell interface for a free-moving outer surface and a fixed outer surface, respectively. The dimensionless parameters a_1 and a_2 incorporate the key physical implications of the boundary conditions.

4 Relative Pressure and Melting Rate Equation

In this section, the key governing equations that couple the elastic properties of the shell with the volume changes produced through the melting process will be described briefly. The total volume change of the PCM produced by the density difference between liquid and solid and the pressure-induced density changes have been previously found [13]. The key thermodynamic parameter used is total mass, and after applying mass conservation to a confined PCM constituted by compressible phases, the total volume change of the PCM is given by

$$\begin{aligned}\Delta V &= \left(\rho_{r_0}\left(1 - p_d K_{l_d}\right) - 1\right) f_s v_{s_0} \\ &\quad - \rho_{r_0} \frac{p_d K_{s_d}\left(1 - p_d K_{l_d}\right)}{1 - p_d K_{s_d}} \left(1 - f_s\right) v_{s_0} - p_d K_{l_d}\left(1 - v_{s_0}\right).\end{aligned} \tag{9}$$

Here, the definition of compressibility has been used to incorporate the density changes induced by the pressure during the melting process. The dimensionless compressibility of phase i, $K_{i_d} = K_i E$, has been defined. In Eq. (9), the volume change of the PCM is given in terms of the relative density at ambient pressure $\rho_{r_0} = \rho_{s_0}/\rho_{l_0}$, the dimensionless relative pressure p_d, the dimensionless compressibility previously defined, the volumetric fraction of melted solid, and the initial volume of the solid phase $v_{s_0} = 4/3\pi r_{s_0}^3$. The volume change of the PCM given through Eq. (9) is equal to the volume change obtained through the radial displacement of the PCM-shell interface. Using the displacements shown in Eqs. (7) and (8), the volume changes in a system with a free outer surface and a fixed outer surface are given by

$$\Delta V = \frac{4}{3}\pi r_0^3 \left((1 + a_1 p_d)^3 - 1\right) \text{ with } \Delta V = \frac{4}{3}\pi r_0^3 \left((1 + a_2 p_d)^3 - 1\right). \tag{10}$$

Here, the dimensionless parameters a_1 and a_2 are given by Eqs. (7) and (8). Using Eqs. (9) and (10), a nonlinear equation for the dimensionless pressure p_d is obtained in each case [13]:

$$\begin{aligned}&\left((\alpha p_d + 1)^3 - 1\right)\left(1 - p_d K_{s_d}\right)\rho_{r_0} p_d K_{s_d}\left(1 - p_d K_{l_d}\right)(1 - f_s)\, v_{s_0}\\ &- \left(\rho_{r_0}\left(1 - p_d K_{l_d}\right) - 1\right)\left(1 - p_d K_{s_d}\right) v_{r_0} f_s \;+ p_d K_{l_d}\left(1 - p_d K_{s_d}\right)\left(1 - v_{r_0}\right) = 0,\end{aligned} \tag{11}$$

where $\alpha = a_1$ and $\alpha = a_2$ correspond to the relative pressure in a system with a free outer boundary and a system with a fixed shell-matrix interface, respectively. Finally, the relative volume of the solid phase, $v_{r_0} = r_{s_0}^3 / r_0^3$, has been defined.

4.1 Melting Rate Equation

In this part of the section, the equation of motion for the solid sphere radius r_s will be described briefly. Applying mass conservation to the system's core and performing a local energy balance at $r = r_s$, the following equation of motion for r_s is obtained [14]:

$$L_f(t)\left(\rho_s(t)\frac{dr_s(t)}{dt} + \frac{1}{3}r_s(t)\frac{d\rho_s(t)}{dt}\right) = \lambda_l \frac{\partial T_l\,(r,t)}{\partial r}\bigg|_{r=r_s} - \lambda_s \frac{\partial T_s\,(r,t)}{\partial r}\bigg|_{r=r_s}, \tag{12}$$

where L_f represents the latent heat at some time t of the melting process. The thermal conductivity of the liquid (solid) phase is represented as $\lambda_\ell(\lambda_s)$. The right-hand side of Eq. (12) is proportional to the thermal flux at $r = r_s$ and depends on the temperature distribution $T_\ell(r,t)$ $(T_s(r,t))$ within the liquid (solid) phase. Equation (12) must be solved numerically for the radius of the solid sphere at any given time during the melting process. In terms of the volumetric fraction of melted solid f_s, using the definition of compressibility, and performing the time derivative of Eq. (11) for the dimensionless pressure [13], the following equation of motion for f_s is obtained:

$$\begin{aligned}&L_f \rho_{s_0} r_{s_0} (1 - f_s)^{-2/3} \frac{df_s}{dt}\left(\frac{h\,(p_d)}{q\,(p_d)} \frac{K_{s_d}}{\left(1 - K_{s_d} p_d\right)^2}(1 - f_s) - \frac{1}{1 - K_{s_d} p_d}\right)\\ &\quad = -\phi\,(r_s, t)\,,\end{aligned} \tag{13}$$

where $\phi(r_s, t)$ is the thermal flux defined in the right-hand side of Eq. (12). This equation is numerically more stable than Eq. (12) when considering the compressibility of both phases. The functions h and q of the dimensionless pressure that appear in Eq. (13) are given by

$$h\,(p_d) = \left(\left(\rho_{r_0} - 1\right) + \left(K_{s_d} - K_{l_d}\rho_{r_0}\right) p_d\right) v_{r_0} \tag{14}$$

and

$$\begin{aligned} q\,(p_d) = {} & (\alpha p_d + 1)^2 \left(3\alpha - K_{s_d}\,(1 + 4\alpha p_d)\right) + \rho_{r_0} v_{r_0} \left(K_{s_d} + K_{\ell_d} f_s - 2K_{s_d} K_{\ell_d} p_d\right) \\ & + \left(1 - v_{r_0}\right)\left(K_{\ell_d} - 2K_{\ell_d} K_{s_d} p_d\right) + K_{s_d}\left(1 - f_s v_{r_0}\right), \end{aligned} \tag{15}$$

where $\alpha = a_1$ or $\alpha = a_2$, depending on the boundary condition applied at the outer surface. Finally, we will describe briefly the steps needed to find the numerical approximations to the volumetric fraction of melted solid and the temperature distribution in each phase. The first step consists of finding the fraction of melted solid at the next time level, through a first-order approximation of the time derivative of f_s in Eq. (13).

The second step consists of finding the only possible root for p_d from the solution of Eq. (11). Regarding the system with a free outer surface, the maximum possible relative pressure can be determined through the incompressible limit [13]. The root that lies in the interval $\left[0,\ p_d^{\max}\right]$ at any given time is chosen as the solution for p_d. Additionally, for a system with a fixed shell-matrix interface, the inner displacement $u_r(r_0) = 0$, when $r_0 \to R_0$, which corresponds to an isochoric transition. In this limit, and $a_2 \to 0$, Eq. (11) is reduced to a second-degree polynomial for p_d. The maximum possible relative pressure is found by substituting $f_s = 1$ in the corresponding solution as follows:

$$p_d^{\max} = \frac{1}{K_{\ell_d}}\left(1 - \frac{1}{\rho_{r_0}}\right). \tag{16}$$

The inner displacement is found by substituting the solution for p_d in Eq. (7) or Eq. (8). The density of the liquid (solid) phase is determined through the definition of compressibility. The liquid-solid saturation properties are obtained through a second-order expansion of the change in chemical potential $\Delta\mu = \mu_\ell - \mu_s$ at the saturation point [8]. Finally, the temperature distribution at the corresponding relative pressure is estimated by solving the heat equation in each phase through a heat balance integral method [13].

5 Results and Discussion

In this section, we will describe the parameters used to estimate the thermal performance of encapsulated PCMs in each case considered: free outer surface and fixed shell-matrix interface. The thermal performance is described through the inner

Table 1 Thermodynamic properties at 1 atm of PCMs considered [8, 15]

Salt	T_m (K)	L_f (kJ/kg)	ρ (kg/m^3)	C_p (kJ/kgK)	λ (W/mK)	K (GPa^{-1})
KNO_3	607	100	1800.89 (liq.)	1.517 (liq.)	0.425 (liq.)	0.269 (liq.)
			1870 (sol.)	1.4 (sol.)	0.5 (sol.)	0.048 (sol.)
$NaNO_3$	594	173	1875.03 (liq.)	1.805 (liq.)	0.516 (liq.)	0.247 (liq.)
			2111.2 (sol.)	1.739 (sol.)	0.5 (sol.)	0.037 (sol.)
KNO_3-	496	105	2096 (liq.)	1.5 (liq.)	0.8 (liq.)	0.186 (liq.)
$NaNO_3$			2192 (sol.)	1.43 (sol.)	1.0 (sol.)	0.031 (sol.)

relative pressure and the elastic limit of spherical shells, the energy density storage during the melting process, and heat transfer rates. Three types of salts confined in SiC shells will be analyzed. Table 1 shows the relevant thermodynamic properties of KNO_3, $NaNO_3$, and the corresponding eutectic $KNO_3/NaNO_3$.

5.1 *Relative Pressure and Elastic Limit*

In this part of the section, results concerning the solution of Eq. (11) will be presented. The relative pressure is found by considering the free outer surface and the fixed outer surface scenarios. The relative pressure where the encapsulating material reaches the elastic limit will be obtained by using the Von Mises criteria [16]. According to this criterion, the dimensionless critical pressure where a spherical shell reaches the elastic limit is given by

$$p_c = \frac{2}{3}\frac{\sigma_Y}{E}\left(1 - \frac{r_c^3}{R^3}\right), \tag{17}$$

where σ_Y is the yield stress of the encapsulating material, r_c is the inner radius of the shell in the elastic limit, and R is the outer radius in this limit. The values of r_c and R can be expressed in terms of the dimensionless pressure through the corresponding displacements $u_r(r_0)$ and $u_r(R_0)$. Considering a free outer surface, the outer radius of the shell can be found by evaluating Eq. (5) at $r = R_0$ [13]. Alternatively, for a fixed outer surface, the outer radius of the shell can be found by evaluating Eq. (6) at $r = R_0$. The later process leads to the following expressions for the outer radius in each case:

$$R = (1 + bp_d) \text{ with } b = \frac{3}{2}\frac{r_0^3}{R_0^3}\left(\frac{1-\nu}{1 - r_0^3/R_0^3}\right) \text{ for a free outer surface, and}$$

$$R = R_0 \text{ for a fixed outer surface.} \tag{18}$$

Applying Eqs. (7), (8), and (18) to the critical pressure shown through Eq. (17) leads to the following expressions for the dimensionless critical pressure:

$$p_c^{(1)} = \frac{2}{3}\frac{\sigma_Y}{E}\left(1 - \frac{r_0^3}{R_0^3}\frac{\left(1 + a_1 p_c^{(1)}\right)^3}{\left(1 + b p_c^{(1)}\right)^3}\right) \text{ and } p_c^{(2)} = \frac{2}{3}\frac{\sigma_Y}{E}\left(1 - \frac{r_0^3}{R_0^3}\right)\left(1 + a_2 p_c^{(2)}\right)^3. \tag{19}$$

Here, $p_c^{(1)}$ and $p_c^{(2)}$ represent the critical pressure for a system with a free outer surface and a system with a fixed shell-matrix interface, respectively. Equations (11) and (19) can be solved simultaneously, to find the critical pressure and the fraction of melted solid where the elastic limit is reached for a given shell's thickness.

Figure 1 shows the relative pressure at full melting fractions $f_s = 1$ for microencapsulated salts initially in their solid phase $r_{s_0} = r_0$. The relative pressure ΔP is shown in terms of the initial shell's relative thickness $\widetilde{\delta_0} = \delta_0 / r_0$, where δ_0 is the initial thickness of the shell. The relative pressure is obtained for the three salts described in Table 1. The relative pressure $\Delta P^{(1)}$ and $\Delta P^{(2)}$ is estimated through the solution of Eq. (11) for a free-moving outer surface and a fixed shell-matrix interface, respectively. The isobar for each salt observed in Fig. 1 corresponds to the relative pressure when the initial relative thickness $\widetilde{\delta_0} \to \infty$. In this limit, both models represent the same physical system of a PCM confined in an infinite medium constituted by the shell's material. The dimensionless parameters $a_1 = a_2$ when $R_0 \gg r_0$. Taking the corresponding limit in Eqs. (7) and (8), the parameters a_1 and a_2 become $a = (1 + \nu)/2$. Additionally, at full melting fractions and for systems that initially are in their solid state, Eq. (11) is reduced to the following expression:

$$\left((ap_d + 1)^3 - 1\right) + p_d \rho_{r_0} K_{\ell_d} - \left(\rho_{r_0} - 1\right) = 0. \tag{20}$$

The last equation has only one real root, which can be found through Cardano's solution, and represents the isobars shown in Fig. 1. Note that both models approach asymptotically to this relative pressure. Young's modulus of $E = 476$ GPa and Poisson's ratio of $\nu = 0.19$ for a silicon carbide shell were considered [9]. Finally, the minimum relative thickness for which the shell reaches the elastic limit exactly at $f_s = 1$ can be found through the solution of Eqs. (11) and (19) for p_c and δ_0. The minimum relative thickness represents a threshold thickness $\widetilde{\delta_t}$ from which the shell will be in the elastic regime at any value of f_s. Table 2 shows the values of $\widetilde{\delta_t}$ and the corresponding maximum relative pressure for each of the salts considered. Additionally, the threshold thickness was calculated for a free outer surface $\widetilde{\delta_t^{(1)}}$ and a fixed shell-matrix interface $\widetilde{\delta_t^{(2)}}$.

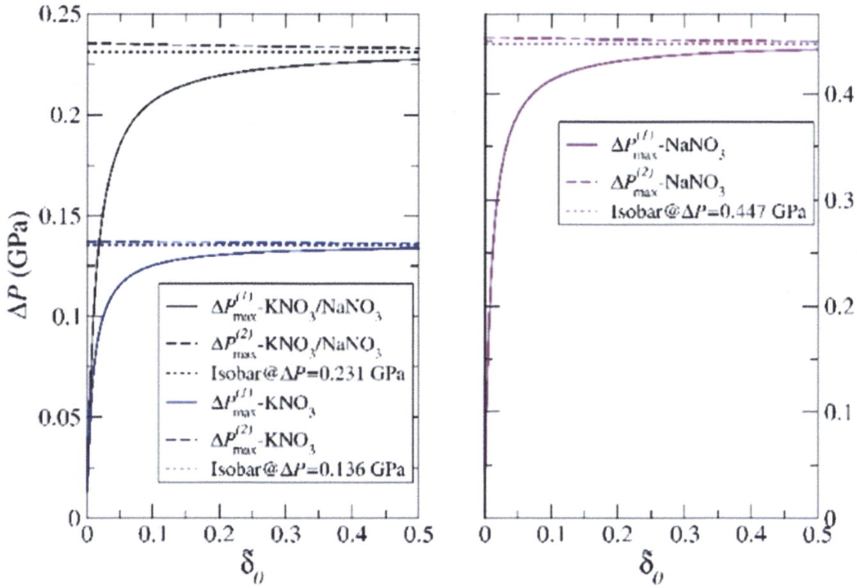

Fig. 1 Relative pressure. ΔP at full melting fractions, in terms of the initial relative thickness of the SiC shell. The horizontal dotted lines represent the isobars where the two models show the same behavior, which is in the limit of infinite thickness

Table 2 Threshold thicknesses

Salt	$\widetilde{\delta_t^{(1)}}$	$p_c^{(1)}$ (GPa)	$\widetilde{\delta_t^{(2)}}$	$p_c^{(2)}$ (GPa)
KNO_3	0.0114	0.0757	0.0208	0.1373
$NaNO_3$	0.0656	0.3954	0.0760	0.4522
KNO_3-	0.0227	0.1484	0.0367	0.2352
$NaNO_3$				

Initial relative thickness and critical pressure for which the SiC shell reaches the elastic limit at full melting fractions. Results are shown for each model considered: (1) free-moving outer surface and (2) fixed shell-matrix interface. A yield stress of $\sigma_Y = 3.44$ GPa for SiC has been used

5.2 *Energy Density and Melting Times of Microencapsulated Salts*

The volumetric energy density is a key parameter that helps to assess the thermal performance of salts used in thermo-solar applications. Energy density represents the energy storage capacity of the system used to meet the energy supply/demand chain. The internal energy of confined PCMs has been obtained previously by considering the compressibility of each phase and the density difference between liquid and solid phases [14]. The main approach consists on using total mass conservation when incorporating the deformation of each phase and system's volumetric expansion during the melting process. The following four stages are considered to find the sensible heat stored in a small-time interval Δt: (a) sensible heat absorbed $\Delta U_\ell^{(1)}$ by the existing mass of liquid at time t; (b) internal energy change $\Delta U_s^{(1)}$ of the solid mass that is not melted during this small-time interval; (c) heat absorbed by the mass of solid $\Delta U_s^{(2)}$ that will be melted, from its initial temperature at time t to the melting temperature T_m of the PCM at the current pressure; and (d) heat absorbed by the melted mass of solid in its liquid form $\Delta U_\ell^{(2)}$, from its melting temperature T_m at the final pressure to its final temperature at $t + \Delta t$.

The thermal energy absorbed ΔQ is found by considering the total amount of sensible heat, obtained by adding the contribution of each stage just described and the contribution of the latent heat stored in a small-time interval. The energy density is estimated through the total thermal energy stored and the volume of the PCM. Figs. 2 and 3 show the energy density stored by each salt where $r_0 = 1$ mm and $r_{s_0} = 0.99\ r_0$, for different relative thicknesses and according to the models considered in this work. Fig. 2a shows the energy density stored by the KNO_3/$NaNO_3$ salt at melting fractions of $f_s = 0.99$. The system with an outer free surface is best behaved for this salt when the relative thickness is close to the threshold value of $\widetilde{\delta_t} = 0.0227$. Here, the energy density of the eutectic salt has a peak value of approximately 454.79 MJ/m^3. The corresponding melting time is 14.83 s. Consequently, this is the optimum configuration for this PCM-shell system.

The energy density observed in Figs. 2c and 3a grows monotonically with the relative thickness for the free outer surface shell system. The behavior is in contrast with latent heat storage in confined systems, where latent heat tends to decrease with the shell's thickness. However, one factor that could explain the observed behavior is the higher compressibility of KNO_3 and $NaNO_3$ salts, which is related with latent heat decrease rate and total core's volume change. Additionally, according to the results illustrated in Figs. 2c and 3a, the system with a fixed shell outer surface shows the best energy storage efficiency. However, melting times are greatly increased for these types of systems, according to the results shown in Figs. 2d and 3b. Consequently, it is more subtle to determine the optimum thermal performance in these cases.

6 Conclusions

The relationship between compressibility and thermal performance in two types of microencapsulated systems has been illustrated in this work. We found that in a eutectic mixture of salts, the optimum energy storage-melting time ratio is obtained in SiC shells with a free outer surface and close to the threshold thickness of $\widetilde{\delta_t} = 0.0227$. In contrast to this salt, the energy density of KNO_3 and $NaNO_3$ salts shows a monotonic growth with relative thickness in free outer surface capsules, while

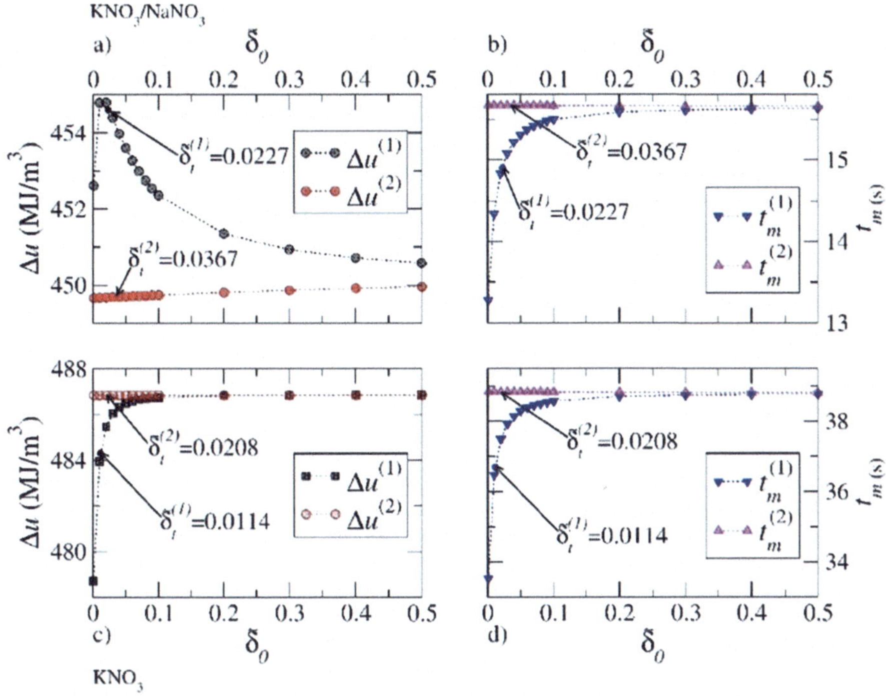

Fig. 2 Microencapsulated salts with an initial core radius of $r_0 = 1$ mm and initial solid radius of $r_{s0} = 0.99$ mm. (**a**) Energy density of the eutectic salt and (**c**) KNO_3. (**b**) Melting times for the eutectic salt and (**d**) the potassium nitride salt KNO_3

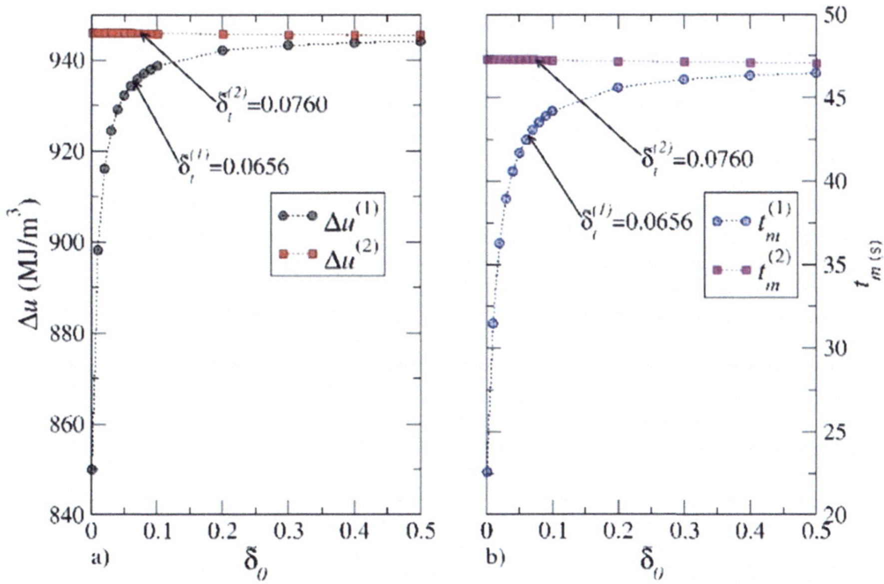

Fig. 3 (**a**) Energy density and (**b**) melting times for a microencapsulated $NaNO_3$ salt

the best energy storage capacity is observed in fixed outer surface systems. The best thermal performance for the KNO_3 and $NaNO_3$ salts is obtained for a free outer shell's surface and when the shell has the largest relative thickness. An energy density to melting time ratios of 12.554 MJ/s m^3 and 20.31 MJ/s m^3 for the KNO_3 and $NaNO_3$ salts is obtained with the present model. The proposed model can be used as a criterion for PCM-shell system selection in thermal energy storage backup systems.

Acknowledgements The authors would like to thank Conahcyt for their financial support through the project number CF-2023-G-792.

References

1. Palacios, A., et al.: Thermal energy storage technologies for concentrated solar power–a review from a materials perspective. Renew. Energy. **156**, 1244–1265 (2020)
2. Chiew, J., et al.: Low-temperature macro-encapsulated phase change material based thermal energy storage system without air void space design. Appl. Therm. Eng. **141**, 928–938 (2018)
3. Li, F., et al.: Application and analysis of flip mechanism in the melting process of a triplex-tube latent heat energy storage unit. Energy Rep. **9**, 3989–4004 (2023)
4. Du, Z., et al.: Numerical studies on a fin-foam composite structure towards improving melting phase change. Int. J. Heat Mass Transf. **208**, 124076 (2023)
5. Shu, G., et al.: Effect of charging/discharging temperatures upon melting and solidification of PCM-metal foam composite in a heat storage tube. Int. J. Heat Mass Transf. **201**, 123555 (2023)
6. Huang, X., et al.: Investigation and optimization of solidification performance of a triplex-tube latent heat thermal energy storage system by rotational mechanism. Appl. Energy. **331**, 120435 (2023)
7. Sarı, A., et al.: Microencapsulated heptadecane with calcium carbonate as thermal conductivity-enhanced phase change material for thermal energy storage. J. Mol. Liq. **328**, 115508 (2021)
8. Lopez, J., et al.: Confined melting in deformable porous media: a first attempt to explain the graphite/salt composites behaviour. Int. J. Heat Mass Transf. **53**(5–6), 1195–1207 (2010)
9. Pitié, F., Zhao, C.Y., Cáceres, G.: Thermo-mechanical analysis of ceramic encapsulated phase-change-material (PCM) particles. Energy Environ. Sci. **4**(6), 2117–2124 (2011)
10. Parrado, C., et al.: Thermo-mechanical analysis of copper-encapsulated NaNO3–KNO3. Chem. Eng. Res. Des. **93**, 224–231 (2015)
11. Yu, Q., et al.: Thermo-mechanical analysis of microcapsules containing phase change materials for cold storage. Appl. Energy. **211**, 1190–1202 (2018)
12. Benner, J.Z., et al.: The effect of micro-encapsulation on thermal characteristics of metallic phase change materials. Appl. Therm. Eng. **207**, 118055 (2022)
13. Otero, J.A., Hernández-Cooper, E.M.: Elastic limit and charging times of compressible microencapsulated phase change materials. Appl. Therm. Eng. **237**, 121764 (2024)
14. Hernández-Cooper, E.M., Otero, J.A.: Effects of pressure-induced density changes in the thermal energy absorbed by a micro-encapsulated phase-change material. Molecules. **24**(7), 1254 (2019)
15. D'Aguanno, B., Karthik, M., Grace, A.N., Floris, A.: Thermostatic properties of nitrate molten salts and their solar and eutectic mixtures. Sci. Rep. **8**(1), 1–15 (2018)
16. Gao, X.L.: Strain gradient plasticity solution for an internally pressurized thick-walled spherical shell of an elastic–plastic material. Mech. Res. Commun. **30**(5), 411–420 (2003)

Energy Storage in the Smart Grid: A Multi-agent Deep Reinforcement Learning Approach

Pawel Knap and Enrico Gerding

1 Introduction

This chapter proposes an energy storage solution controlled by Deep Reinforcement Learning (DRL) to address fluctuating electricity costs in the smart grid (SG). Utilizing real-world data from the Low Carbon London project [1] and Octopus variable tariff data [2], a self-interested DRL agent makes decisions based on price, storage level, and stored electricity value every 30 minutes. The study investigates the concurrent usage of storage and photovoltaic (PV) panels and simulates a community of households to evaluate their behaviour, cooperation–competition patterns, and impact on the power grid. Various agent types, action capabilities, storage capacities, and PV powers are tested. Results indicate significant consumer savings and grid stress reduction. In summary, our study examines the benefits and challenges of SG, highlighting the effectiveness of in-house energy storage controlled by a selfish DRL agent. Full source code and supplementary material are available at https://github.com/PawelKnap/EneStore.

2 Background

In reinforcement learning (RL), an agent aims to optimize its policy to maximize cumulative rewards. The RL cycle, shown in Fig. 1, involves the agent selecting actions, thus altering the environment state and receiving rewards. A policy $\pi(a_1, s_1) = Pr(a_1|s_1)$ dictates the agent's actions based on the current environment

P. Knap (✉) · E. Gerding
University of Southampton, Southampton, UK
e-mail: pmk1g20@soton.ac.uk

L. Chen (ed.), *Trends in Clean Energy Research*, Green Energy and Technology,
https://doi.org/10.1007/978-3-031-67987-2_21

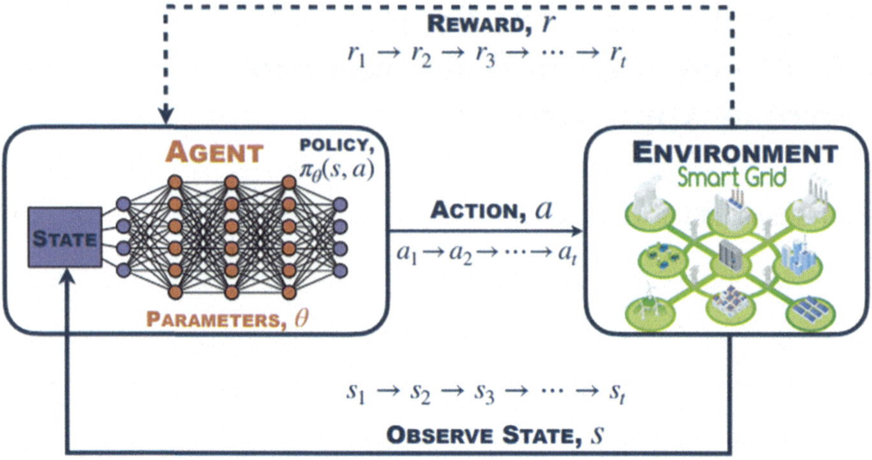

Fig. 1 Reinforcement learning cycle. An agent in state s_1 chooses action a_1, affecting the environment to transition to s_2 and receiving reward r_1, and the whole process repeats itself (Adapted from [3])

state. The value function is computed as $V(s_1) = E(\sum_{t=1}^{\infty} \gamma^t r_t | s_1)$, approximating the expected cumulative reward in a given state. Another key concept is the quality function, $Q(s_1, a_1) = E[R(s_2, s_1, a_1) + \xi V(s_2)]$, assessing state–action pairs' quality considering immediate and future rewards. This function operates within a Markov Decision Process (MDP). MDP's crucial feature is the conditional independence of the next state from prior states and actions. The value and policy functions can be derived from the Q function, with the former representing maximum value as $V(s_1) = max_{\mathrm{a}} Q(s_1, a)$ and the latter indicating the action with the maximum $V(s_1)$ as $\pi(a_1, s_1) = argmax_{\mathrm{a}} Q(s_1, a)$.

2.1 Used Method Overview

In Q-Learning, an agent updates its Q function iteratively using the formula: $Q^{new}(s_1, a_1) = Q^{old}(s_1, a_1) + \alpha[r_1 + \gamma max_a Q(s_2, a) - Q^{old}(s_1, a_1)]$, where α is the learning rate. Here, $r_1 + \gamma max_a Q(s_2, a)$ represents the target estimate of cumulative reward, serving as part of an error signal which guides the Q function update, with zero error indicating optimal policy discovery. The update increases the Q value if the agent's reward exceeds expectations or decreases it otherwise. In practice, the Q function is a table of Q values for different states and actions.

Deep Q-Learning (DQL) [4], an off-policy gradient-free technique, employs deep neural networks (NNs) to approximate the Q function for decision-making. While very deep NNs excel in tasks like Computer Vision [5], DRL often opts for simpler NN architectures with 2–3 hidden layers [6, 7]. During training, it balances exploration and exploitation through trial and error, gradually transitioning from random to informed actions based on experience, focusing exploration on promising paths. In DQL, the quality function $Q(s_1, a_1)$ is parameterized by NN weights Θ, addressing high-dimensional feature approximation and the curse of dimensionality [8]. The NN minimizes the squared error expectation via gradient descent and backpropagation to optimize parameters Θ, determining the optimal Q function. Experience replay [9] enhances generalization using previous transitions to update NN weights. Additionally, two NNs, the policy and target network, manage correlations and improve stability.

3 Related Work

While there is agreement that storage presence in SG is beneficial, optimal deployment location remains debated. Barbour et al. [10] advocate for community batteries over individual household storage to reduce energy exchange with the grid. However, their approach increases demand peaks, adding to grid stress. Voice et al. [11] propose decentralized control of micro-storage selfish agents in a Multi-Agent System (MAS), reducing supplier costs by 16% and enhancing robustness, yet constant daily load and equal import–export price assumptions may limit applicability. Their approach also overlooks the self-use of stored energy and renewable integration. Wang et al. [12] suggest shared ownership of household storage, improving peak reduction and network investment, but our approach offers similar advantages while being simpler. Furthermore, their joint storage control raises privacy concerns. Deng et al. [13] propose price schemes to influence consumption patterns, thus reducing demand peaks, whereas our solution achieves the same without requiring behaviour changes.

An overview of energy marketplace models and dynamic pricing techniques for SG is provided in [14–16]. A recent study [17] introduces an RL-based energy market model for prosumer-dominated microgrids. Employing multi-agent reinforcement learning (MARL), it establishes a dynamic pricing environment linked to real-time demand, resulting in increased profits for prosumers and the grid operator, along with reduced grid reserve power utilization. However, it focuses solely on prosumers, neglecting consumers. Another work [18] presents a MARL-based solution for industrial sites to manage electricity costs amid fluctuating prices and growing renewable power generation. This approach optimizes production resources, battery storage, self-generation, and market trading to minimize expenses. Comparative assessments demonstrate the superiority of the MARL system over rule-based strategies in terms of speed and quality. However, it simplifies cooperative agents' behaviours by considering only essential information, potentially overlooking their complete intricacy.

Machine Learning (ML) algorithms have diverse applications in SG. For example, Atef and Eltawil [19] forecast electricity prices, Asare-Bediako et al. [20] use NNs for residential load profile forecasting, and Eck et al. [21] predict local energy demand. Furthermore, Singh et al. [22] propose a novel MAS-based system for load frequency control, outperforming previous algorithms, leveraging distributed RL controllers and swarm optimization. Additionally, Qiu et al. [23] explore DRL's approach to peer-to-peer energy trading, offering an alternative SG functioning method. Ali and Choi [24] provide a comprehensive review of ML techniques in SG, highlighting six main development areas and discussing market liberalization and economic aspects.

4 Experimental Setup

The experiment used electricity consumption data from the Low Carbon London project [1], involving 5,567 London households' smart meters data from November 2011 to February 2014. This data was merged with variable tariff prices from Octopus Energy [2], resulting in a dataset spanning over 15 million episodes for single-agent simulations. Storage sizes of 0.5 kWh, 1.5 kWh, and 3 kWh were studied, corresponding to available in-house socket powers. Agent operations are shown in Fig. 1. Agents make decisions based on the environmental state in each half-hour episode, with actions including waiting, purchasing, using, or selling energy. The range of environment states varies depending on the scenario, from 110 in single agent to infinity in MAS simulations.

4.1 Simulation of Solar Panels Energy Generation

To model households as prosumers, we employed an algorithm simulating PV energy generation due to insufficient real data for the long training period. For each interval excluding nighttime, energy production in kWh was computed. Daytime duration, average sunrise, and sunset times were calculated for London, remaining constant for 30 days, resulting in a 360-day simulation year. Solar power production patterns were simulated using Gaussian functions with small random variations. Additionally, brief periods of low energy production due to cloud cover were simulated by adjusting energy production with probabilities. Figure 2c illustrates a typical solar generation pattern. The energy within each interval is scaled based on the average daily energy production, which varies monthly. Simulations can be repeated for any number of years, with each daily generation pattern being unique. The simulated annual energy generation varies between 800 and 900 kWh for a 1 kW system (proportionally more for higher powers), as detailed in [25].

4.2 Single-Agent Simulation

4.2.1 Baseline

Algorithm operates on a simple rule: It buys energy for storage when the price is below 9 p/kWh, uses stored energy when the price exceeds 18 p/kWh, and waits otherwise. Any electricity deficit is bought from the grid. In the prosumer simulation, all generated energy is immediately sold.

4.2.2 Deep Q-Learning Agent

Employs the reward policy detailed in Algorithm 1. Epsilon linearly decreases from 1 to 0 over 15 million training examples, while gamma is set to 0. A replay memory of 1024 episodes updates a neural network with one hidden layer of 2048 units, using a batch size of 32 examples and the Adam optimizer with a learning rate of 10^{-9}. The target network updates every 1000 episodes. Similar to the baseline, the agent chooses from three actions (buy, use, and wait) in each time interval, purchasing any deficit from the grid. The agent's environment state is defined by storage level, stored energy value, and current electricity price. To optimize results, the latter two variables are binned into ten categories based on price per kWh, while storage filling is binary (empty or not), resulting in 110 states due to the correlation between storage filling level and stored energy value (which is 0 when storage is empty).

4.2.3 DQL Agent with Increased Action Space

Exploring the addition of a fourth action allowing agents to sell stored energy aimed to boost savings in scenarios with full storage, high export prices, and low energy consumption or when export prices exceed import prices. The NN incorporates the selling price as a fourth input, modelled by the Octopus export tariff binned similarly to import prices (10 bins each). Thus, the state space increases from 110 to 1,100. The reward policy, including the red colour code, is detailed in Algorithm 1. All other parameters remain unchanged from Sect. 4.2.2.

4.2.4 DQL Agent as a Single Prosumer

Is created by integrating the PV simulator into the 3-action DQL agent without retraining. Octopus's export tariff determines the sell price. The produced energy is firstly used for household consumption, with surplus either stored or sold if storage is full. Alternatively, generated energy is added to storage or sold if storage space is insufficient. The former excelled in single-agent simulations, while the latter performed better in MAS simulations, as explained in the Appendix.

4.3 MAS Simulation

3-action DQL agents were combined in a MAS to simulate a community of three households, with each represented by one retrained agent. They operated in a shared electricity market with prices determined by community demand. Various price–demand functions, including linear, logarithmic, and exponential, were modelled as follows: $Price = 3 \times demand$, $Price = 11 \times ln(3 \times demand + 0.7)$, $Price = exp(demand/11.25) - 1$. The decision to include the logarithmic function was influenced by findings from Lipman [26], showing a logarithmic relationship between net demand and price in the Octopus tariff. Agents underwent retraining with 1.2 million episodes in the MAS scenario, maintaining a reward policy as described in Sect. 4.2.2. Threshold price values for purchasing decisions were 3.075, 9.04, and 0.19 p/kWh for linear, logarithmic, and exponential functions. They were chosen based on the observation that around 60% of instances in the training set had prices lower than these thresholds. The same principle guided the setting of the original threshold values.

Agent's NNs have two hidden 2048-wide layers, with continuous input data on price, stored energy quantity, and its value. Distinct 128-wide replay memories provide training batches of 32 samples, with target networks updated every 100 episodes. Adam optimizers with a learning rate of 10^{-6} were employed. Gamma was set to 0, and epsilon decreased linearly from 1 to 0. Testing was conducted on three datasets, each with 300,000 episodes per agent. In the MAS with the logarithmic price function, an adjustment was made to accommodate negative prices, with such purchases receiving a reward of 10. Since the initial price depends on community demand, it is assumed the energy supplier accurately predicts demand for the upcoming interval. However, agents' actions influencing demand cause price variation, posing a challenge as they base decisions on the initial price, different from the final price used for bill calculations. Finally, PV data was included in the MAS simulation with logarithmic prices to model households as prosumers. The selling price is set at 80% of the import price, unless the import price is negative, in which case the selling price is 0, effectively mirroring real-world scenarios due to inherent regularization.

5 Results

5.1 Octopus Agile Import and Export Tariff Savings

Switching from a fixed tariff to the variable Octopus Agile tariff can yield savings for households, as shown in Table 4. Costs rise only in 1 out of 40 cases due to high peak consumption, but our storage system offsets this loss. Unless specified, all simulations measure savings relative to the Octopus Agile tariff's electricity cost. In the first quarter of 2019, the mid-level PV export tariff was 3.54 p/kWh, according to Ofgem [27]. This rate is used here to demonstrate the advantages of a variable export tariff. Table 5 presents annual profit statistics resulting from the tariff change, compared to earnings under the Octopus Agile export tariff.

5.2 Results of a Single-Agent Simulations

Table 1 illustrates the superiority of the DQL approach over the baseline, with a slight improvement for the 4-action DQL, except in the case of a 0.5 kWh storage. The simulations underscore that significant savings stem from combining storage and solar panels, showcasing agents' efficacy as prosumers. However, expanding battery capacity for a given PV power does not substantially increase savings, as evaluated battery sizes are not adequate for simulated PV power generation.

5.3 Results of MAS Simulations

Table 2 demonstrates varying savings influenced by individual agents' consumption patterns. The exponential function produces the highest savings, followed by the linear function, indicating that steeper price–demand functions correlate with greater savings. Additionally, specific consumption patterns contribute to enhanced savings, with agent 2 achieving the highest savings, with agent 0 ranking second across all price–demand functions. Furthermore, Table 3 displays saving rates for simulations of MAS with solar panels' energy generation.

Not all consumers may prefer to use storage. Hence, the analysis extends to evaluate the impact of households with storage on the electricity expenses of those without it. The resulting cost reductions for various scenarios and the percentage deviation in savings compared to the scenario where all agents have storage facilities are summarized in Table 6. This table also highlights that agents exert varying influences on neighbourhood savings. The absence of storage in agent 0 slightly affects the savings of all agents, including itself. Agents 1 and 2 experience a decrease of under 3% in scenarios where other agents lack storage. However, if they are without storage, their savings drop by more than 50%. Notably, agent 0 sees a significant decline in savings when any community member loses access to storage. Nevertheless, each agent saves when at least one of them possesses a storage system.

6 Discussion

6.1 Single-Agent Simulations

Figure 2a illustrates a logarithmic increase in median saving rates with storage size, plateauing around 22% for capacities over 4 kWh. These findings, coupled with the significantly higher costs of high-capacity storage, make the installation of large storage facilities economically unfeasible. Technological limitations also arise, such as the maximum power capacity of home charging points. Furthermore, analysis shows savings decrease as total yearly electricity consumption increases (Fig. 2b).

Table 1 Mean yearly savings from various agents with different storage sizes, relative to no storage case. With PV, surplus energy is immediately sold

	0.5 kWh battery		1.5 kWh battery		3 kWh battery	
Metrics	Savings in £	Savings in %	Savings in £	Savings in %	Savings in £	Savings in %
Baseline	19.26 ± 2.81	6.35 ± 2.79	44.97 ± 12.95	13.39 ± 3.67	61.27 ± 22.2	17.42 ± 3.77
DQL	21.7 ± 3.01	7.13 ± 3.08	51.4 ± 14.92	15.21 ± 4.01	68.91 ± 26.96	19.2 ± 3.76
DQL 4-actions	21.63 ± 2.77	7.01 ± 3.31	51.89 ± 13.31	15.35 ± 5.03	73.23 ± 23.6	20.91 ± 5.76
DQL with 1 kW PV	74.17 ± 23.86	24.97 ± 13.62	77.19 ± 24.79	25.94 ± 13.82	78.4 ± 25.94	26.49 ± 14.72
DQL with 4 kW PV	275.39 ± 74.72	90.33 ± 45.41	275.53 ± 74.76	90.27 ± 45.4	275.93 ± 74.96	90.47 ± 45.38

Table 2 Savings as a percentage ratio between costs with and without the 0.5 kWh storage system under different price–demand functions in MAS simulation

Price–demand function	Agent 0 savings in %	Agent 1 savings in %	Agent 2 savings in %
Logarithmic	1.87 ± 0.56	1.52 ± 0.35	2.52 ± 0.24
Linear	3.16 ± 0.45	2.73 ± 0.38	4.75 ± 0.29
Exponential	5.05 ± 1.57	4.65 ± 0.75	5.67 ± 0.4

Table 3 Savings expressed as a percentage ratio between costs with storage and PV versus costs when all agents are without storage, in parentheses without storage and PV. Agents operate in a shared market with logarithmic prices. Cases 1, 2, and 3 involve all agents having storage of 0.5, 1.5, and 3 kWh, coupled with PV of 1, 4, and 6 kW, respectively. In case 4, agent 0 has 0.5 kWh storage and 1 kW PV, agent 1 has 1.5 kWh storage and 4 kW PV, and agent 2 has 3 kWh storage and 6 kW PV. In case 5, agents 0 and 1 lack both storage and PV, while agent 2 has 3 kWh storage and 6 kW PV

Case	Agent 0 savings in %	Agent 1 savings in %	Agent 2 savings in %
1	2.11 ± 0.57 (24.34 ± 2.69)	0.89 ± 0.53 (21.02 ± 3.69)	2.77 ± 1.13 (21.82 ± 1.8)
2	2.04 ± 0.78 (43.39 ± 2.33)	1.07 ± 0.83 (36.27 ± 6.28)	4.95 ± 0.99 (38.59 ± 1.75)
3	1.87 ± 0.9 (46.06 ± 1.83)	-0.2 ± 0.21 (37.85 ± 6.09)	4.88 ± 0.65 (41.3 ± 2.44)
4	2.45 ± 1.32 (40.95 ± 2.83)	1.52 ± 0.78 (35.81 ± 6.14)	9.99 ± 1.11 (41.54 ± 0.87)
5	1.28 ± 0.67 (28.57 ± 2.6)	0.83 ± 0.31 (23.95 ± 4.1)	9.44 ± 1.97 (38.11 ± 0.63)

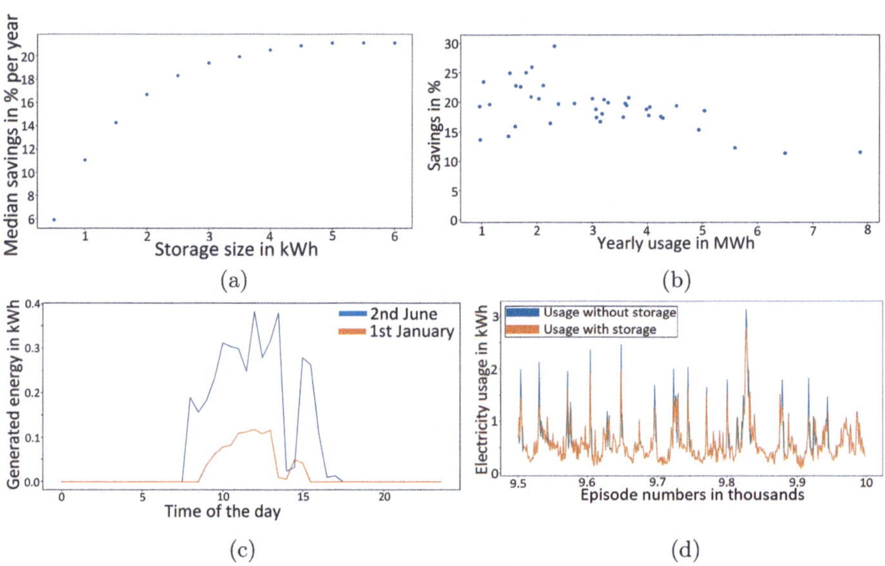

Fig. 2 Median percentage yearly saving of DQL agent versus storage capacities showed in (**a**) and a comparison of yearly usage and savings of DQL agent for different households each using 3 kWh storage in (**b**). Subfigure (**c**) displays solar energy generation patterns for June 2 (blue) and January 1 (orange) generated by our simulation for a 1 kW PV. Subfigure (**d**) compares community demand without (blue) and with (orange) 0.5 kWh storage for all agents

6.2 MAS Simulation

The MAS simulation yielded significant findings. First, certain functions, notably the exponential one, offer more benefits to consumers (Table 2). This is because agents utilize stored energy during peak times and buy it during off-peak hours. The steeper price–demand function results in higher price disparities between peak and off-peak periods, leading to greater savings. Second, the presence of at least one household with a storage system reduces electricity costs for others without it. However, savings are higher for these households when each has its own storage (Table 3). Third, the agents' behaviour examination reveals no clear signs of cooperation or competition.

Table 3 illustrates that larger battery capacities and more powerful PV systems lead to increased savings compared to households without panels and storage. The values in the table represent savings resulting from all storage within the system. This explains why agents 0 and 1 show non-zero values in case 5, despite lacking storage themselves, and why agent 1 records negative savings in case 3 due to other agents' storage impact. Reducing storage size and PV power for agents 0 and 1 in case 4 does not substantially decrease total savings compared to case 3, mainly due to agent 2's influence. This setup underscores that individual agents' savings hinge on the entire system's behaviour, and even households without storage or PV benefit when neighbours adopt these technologies.

7 Conclusion

In summary, our agent-controlled energy storage system benefits both consumers and suppliers, addressing the challenges of variable tariffs and contributing to SG development. Notably, all agents, even those initially disadvantaged, benefit thus fostering social acceptance of SG. Furthermore, the system smoothens demand curves, minimizing grid load fluctuations.

Key findings include higher consumption correlates with smaller savings, increased storage capacity leads to greater bill reduction, plateauing around 4kWh, and significant savings result from combining the system with PV panels. Single-agent simulations favour immediate self-use, surplus storage, and excess energy sales, while MAS simulations prefer direct storage and surplus sale. The DQL agent maintains a full storage state most of the time, extending the battery lifespan. Octopus Agile price–demand patterns resemble a logarithmic function, with exponential pricing computations yielding optimal savings for MAS. Future work may involve refining the simulation to incorporate real-world dynamics, exploring larger agent populations, and adapting to current consumption patterns.

Appendix

Experimental Setup Details

Algorithm 1 DQL reward policy

```
if action == hold then
    if Price (p/kWh) ≥ 10.71 & storage not empty & Value of energy in storage (p/kWh) ≥ Price (p/kWh) then
        Reward = 1
    else
        Reward = 0.001
    end if
else if action == buy then
    if Storage = empty & Price (p/kWh) < 10.71 then
        Reward = 1
    else if Price per kWh ≥ 10.71 then
        Reward = 0
    else
        Reward = 0.3
    end if
else if action == use then
    if Price (p/kWh) ≥ Value of energy in storage (p/kWh) & Storage not empty & Import price (p/kWh) ≥ Export price (p/kWh) then
        Reward = 1
    else
        Reward = −0.3
    end if
else if action == sell then
    if Export price (p/kWh) ≥ Value of energy in storage (p/kWh) & Export price (p/kWh) ≥ Import price (p/kWh) & Storage not empty then
        Reward = 1
    else
        Reward = −0.3
    end if
end if
```

Simulation of PV Energy Generation Details

Furthermore, real-world data have brief periods of low energy production due to cloud cover. To simulate that, the amount of energy produced in an interval chosen with a 15% probability is multiplied by 0.1 if the previous interval was not cloudy. Otherwise, the probability increases to 60%, creating longer periods of overcast weather.

The average monthly energy production is estimated using the widely available website https://www.renewables.ninja/. These values are then divided by 30 to estimate the daily energy production for each month.

Explanation of Inherent Regularization in MAS with Prosumers

The solar energy generation data are included in the MAS simulation with a logarithmic price scheme to model some households as prosumers. The selling price is established at 80% of the import price, unless the import price is negative, in which case the selling price is set to 0. This mirrors the import–export price relationship seen in the Octopus Agile tariff. This simplified relationship effectively simulates real-world scenarios due to inherent regularization. In our simulation, high export prices would require high import prices, which usually occur during high community demand. However, in such cases, agents seldom have surplus energy to sell, making the impact of high export prices negligible in the simulation.

Result Details

See Tables 4, 5, and 6.

Table 4 Yearly savings resulting from using Octopus Agile tariff versus a constant tariff of 14.228 pence/kWh. Columns are independent (maximum values in £ and % may not come from the same test dataset)

Metrics	Savings in £	Saving as % of total bill
Median	42.78	12.34
Mean	60.78	13.65
Standard deviation	59.77	9
Maximum value	263.61	35.69
Minimum value	−14.02	−3

Table 5 Difference of profits per year between variable and flat (3.54 pence/kWh) export tariffs on solar energy generation simulations with different powers used for single agent investigation. The columns are independent

	1 kW installation		4 kW installation		6 kW installation	
Metrics	Profit in £	Profit in %	Profit in £	Profit in %	Profit in £	Profit in %
Median	42.1	139.61	171.63	138.96	257.54	139.16
Mean	35.75	118.63	147.74	118.61	221.21	118.47
Standard deviation	13.37	44.05	55.21	44.23	81.96	43.7
Maximum value	52.52	175.21	216.35	173.35	324.85	175.99
Minimum value	9.54	30.99	39.98	31.64	57.56	31.05

Table 6 Savings showed as a percentage ratio between costs with and without storage for agents in columns 1–3. Columns 4–7 display the percentage change in savings between cases when all and not all agents have storage. All agents use 0.5 kWh storage in a shared market with a linear price–demand function

	Savings (%)			Savings change (%)		
Agents with battery	Agent 0	Agent 1	Agent 2	Agent 0	Agent 1	Agent 2
0, 1, and 2	3.16 ± 0.45	2.73 ± 0.38	4.75 ± 0.29	–	–	–
1 and 2	2.88 ± 0.27	2.68 ± 0.39	4.66 ± 0.20	−8.86	−1.83	−1.89
0 and 2	1.73 ± 0.54	1.27 ± 0.08	4.72 ± 0.10	−45.25	−53.48	−0.63
0 and 1	2.27 ± 0.28	2.70 ± 0.74	2.08 ± 0.51	−28.16	−1.10	−56.21
2	1.42 ± 0.36	1.21 ± 0.07	4.63 ± 0.17	−44.94	−55.68	−2.53
1	1.98 ± 0.19	2.66 ± 0.74	1.95 ± 0.44	−37.34	−2.56	−58.95
0	0.32 ± 0.20	0.06 ± 0.01	0.13 ± 0.07	−89.87	−97.80	−97.26

Table 7 Frequency of a given action occurrence for different agents in % in the test dataset. An action is unsuccessful when the agent attempts to buy energy with full storage or use electricity with an empty battery. Successful buy or use actions are when energy is actually bought or used

Agent number	Wait action	Successful buy action	Successful use action	Unsucc. buy action	Unsucc. use action
0	46.16	0.09	0.13	53.63	0
1	41.05	3.57	5.44	39.91	10
2	16.35	2.12	2.89	78.65	0

Discussion Details

Recall a household initially incurring a £14.02 loss upon switching to a variable tariff. However, employing the 3-action DQL agent results in substantial savings: £21.24 (4.41%) for 0.5 kWh, £62.35 (12.93%) for 1.5 kWh, and £96.3 (19.98%) for 3 kWh storage. Compared to flat tariff costs, this equates to savings of 1.69%, 11.3%, and 19.25% for the respective storage capacities. This indicates that even households potentially affected by variable tariffs in the SG can benefit from adopting the proposed storage system.

When analyzing an agent's behaviour, it is crucial to distinguish between successful and unsuccessful actions. An action is deemed unsuccessful when the agent tries to purchase energy with full storage or utilize electricity with an empty battery. This unsuccessful action is comparable to waiting, as it does not affect the environment. For detailed statistics on the frequency of action selection in the test dataset, please consult Table 7.

A detailed examination of agents' behaviour reveals no clear signs of cooperation or competition. Agents seem to recognize the influence of others on prices, with instances of more than one agent successfully purchasing or using energy being infrequent, constituting only 1.21% and 1.77% of the total amount of actions, respectively. These cases, where significant differences between initial and final prices occur, could be interpreted as competitive behaviours. Conversely, scenarios where one agent successfully buys, another successfully uses, and a third waits result in minimal price changes, suggesting potential cooperative behaviour. However, such cases are rare, amounting to only 0.05% of total episodes. These findings indicate that selfish agents do not establish implicit relationships with each other.

The difference between the best strategies for generated energy utilization in single-agent and MAS simulations can be attributed to the price–demand relationship in MAS. Storing generated energy and using it only during favourable moments, as observed in the MAS agents' approach, prove more effective in maximizing savings. In MAS, optimal savings are achieved when electricity generation moderately exceeds individual agent demand, highlighting the importance of maintaining export price viability without driving it down. This assumes other agents do not adopt the same strategy, which could impact profitability.

References

1. Schofield, J.R., Carmichael, R., Tindemans, S., et al.: Low Carbon London Project: Data from the dynamic time-of-use electricity pricing trial. UK Data Service, SN (2015)
2. Octopus Energy Group: Octopus agile tariff (November 2022)
3. Brunton, S.L., Kutz, J.N.: Data Driven Science and Engineering Machine Learning, Dynamical Systems, and Control. Cambridge University Press, Cambridge (2021)
4. Mnih, V., Kavukcuoglu, K., Silver, D., Graves, A., Antonoglou, I., Wierstra, D., Riedmiller, M.: Playing Atari with deep reinforcement learning (2013)
5. Hassaballah, A.: Deep Learning in Computer Vision: Principles and Applications. CRC Press, Boca Raton (2020)
6. Song, X., Tu, S., Neyshabur, B.: Observational overfitting in reinforcement learning. In: International Conference on Learning Representations (2020)
7. Yarats, Z., Kostrikov, A., Pineau, F.: Improving sample efficiency in model-free reinforcement learning from images. In: CoRR (2019)
8. Bellman, R.: Dynamic Programming. Dover Publications, Mineola (1957)
9. Lin, L.J.: Reinforcement Learning for Robots Using Neural Networks. Ph.D. thesis, Carnegie Mellon University (1992)
10. Barbour, E., Parra, D., Awwad, Z., González, M.C.: Community energy storage: A smart choice for the smart grid? Appl. Energy **212**, 489–497 (2018)
11. Voice, T., Vytelingum, P., Ramchurn, S., Rogers, A., Jennings, N.: Decentralised Control of Micro-storage in the Smart Grid. AAAI, Washington (2011)
12. Wang, Z., Li, F., Sun, H.: Active demand response using shared energy storage for household energy management. IEEE Trans. Smart Grid **4**(4), 1888–1897 (2013)
13. Deng, Y., Chow, C.: A survey on demand response in smart grids: mathematical models and approaches. IEEE Trans. Ind. Inf. **11**(3), 570–582 (2015)
14. Khoshjahan, M., Soleimani, M., Kezunovic, M.: Optimal participation of PEV charging stations integrated with smart buildings in the wholesale energy and reserve markets. In: IEEE Power Energy Society Innovative Smart Grid Technologies (2020)
15. Bayram, et al.: A survey on energy trading in smart grid. In: GlobalSIP (2014)
16. Khan, A.R., Mahmood, A., Safdar, A., Khan, Z.A.: Load forecasting, dynamic pricing and DSM in smart grid: A review. Renewable Sustainable Energy Rev. **54**, 1311–1322 (2016)
17. Ghasemi, A., et al.: A multi-agent deep reinforcement learning approach for a distributed energy marketplace in smart grids. In: SmartGridComm (2020)

18. Roesch, M., Linder, C., Zimmermann, R., Rudolf, A., Hohmann, A., Reinhart, G.: Smart grid for industry using multi-agent reinforcement learning. Appl. Sci. **10**(19), 6900 (2020)
19. Atef, S., Eltawil, A.: A comparative study using deep learning and support vector regression for electricity price forecasting in smart grids. In: ICIEA (2019)
20. Asare-Bediako, B., Kling, W.L., Ribeiro, P.F.: Day-ahead residential load forecasting with artificial neural networks using smart meter data. In: IEEE Grenoble Conference (2013)
21. Eck, B., Fusco, F., Gormally, R., Purcell, M., Tirupathi, S.: AI modelling and time-series forecasting systems for trading energy flexibility in distribution grids. In: ACM e-Energy (2019)
22. Singh, V.P., et al.: Distributed multi-agent system-based load frequency control for multi-area power system in smart grid. Trans. Ind. Electron. **64**(6), 5151–5160 (2017)
23. Qiu, Y., et al.: Scalable coordinated management of peer-to-peer energy trading: A multi-cluster deep reinforcement learning approach. Appl. Energy **292**, 116940 (2021)
24. Ali, S.S., Choi, B.J.: State-of-the-art artificial intelligence techniques for distributed smart grids: a review. Electronics **9**(6), 1030 (2020)
25. Department for Business, Energy & Industrial Strategy: "Energy Trends: UK renewables" (January 2023)
26. Guy Lipman: Forecasting UK electricity prices. Medium blog post (July 2020)
27. Ofgem: "Feed-in Tariffs (FIT)" (March 2023)

Part IV
Carbon Capture and Sustainable Development

Exploration on the Key Factors to the Successful Implementation of Carbon Capture and Storage (CCS) Projects Based on a CCS Project Inventory

Xinyi Duan

1 Introduction

As climate change such as rising temperature and ocean acidification is spreading all over the world, it has become a pressing problem of the Earth. Climate change is a true urgent issue since it may cause environmental consequences by decreasing biodiversity or destructing habitat. Moreover, there are irreversible changes such as melting of glaciers too. In order to meet the challenge, humans have to take actions immediately since climate change is time-sensitive. Thus, CCS can be a significant technology to improve the condition. By capturing and storing, CCS can either reduce or directly remove a large amount of carbon dioxide that is released by heavy industries from reaching the atmosphere efficiently, slowing down climate change.

In recent years, carbon capture and storage (CCS) has been proposed as a potential method to tackle the continued emissions from fossil-fuels while preventing emissions of CO_2 from reaching the atmosphere [1]. As such, a deeper understanding on the status and challenges of CCS projects will be crucial. However, existing research on the successful implementation of CCS projects is still limited.

A number of carbon capture technologies such as oxyfuel combustion have been developed maturely. The large-scale deployment of carbon capture and storage (CCS) is becoming increasingly urgent in the global path toward net-zero emissions [2], and researches suggest that carbon capture and storage can offer economic and environmental benefits in the transition to net-zero energy systems beyond greenhouse gas emission mitigation [3]. However, the slow pace of CCS implementation is concerning and is due, in part, to lack of financial incentives

X. Duan (✉)
Beijing National Day School, Beijing, China

L. Chen (ed.), *Trends in Clean Energy Research*, Green Energy and Technology,
https://doi.org/10.1007/978-3-031-67987-2_22

for CO_2 geological storage, low public acceptance, lack of consistent government energy policy and CCS regulations, and high capital investment [4].

As such, in this study, I examine the database of existing and planned CCS projects worldwide and analyze the key factors in their successful implementation, as well as the main barriers for CCS. This examination based on historical records could inform investment, management, and implementation of CCS projects to learn from lessons in the past.

2 Methodology

2.1 Data Source

All analyses in this paper are based on the carbon capture, utilization, and storage (CCUS) global database provided by the International Energy Agency (IEA) [5]. This dataset could be obtained at https://www.iea.org/data-and-statistics/data-product/ccus-projects-database. It covers 502 CO_2 capture, transport, storage, and utilization projects worldwide that have been commissioned since the 1970s and have an announced capacity of more than 100,000 t per year (or 1000 t per year for direct air capture facilities), providing information about them such as company, overall status, location, project summary, and cost. In addition, this database also includes project summary and other information of these projects, providing key details that help understand their operation or failure.

3 Data Transformation and Analysis

The CCUS dataset includes the cost information of each project and the capture and storage amount of emissions. I first transformed the cost information all to US current dollars, as they are of different currency. All the exchange rates are provided by the website of Google Finance [6]. The calculation was done by the chart automatically after inputting the exchange rate. For emission capture and storage amount, the data includes different units, including barrels per day, tons per day, cubic foot, and megawatt. I conducted transformation of all the units to tons per day. It is assumed that 85% of CO_2 emitted was captured, based on the reported capture and storage efficiency of ConocoPhillips, a project in Taxus, USA. The transformation of the unit lb./MW to the unit tons is based on the calculation method from the Environmental Protection Agency website [7]. Based on this information, the CO_2 emission rate is 884.2 lb. per MW. Based on one project in this category, the Ordos Basin Project, the capture rate is also 85% of CO_2 emitted. We then assumed the same rate for all the CCS projects using this unit. By multiplying the CO_2 emission rate (884.2), the capture rate (85%) of CO_2 emitted, and the unit transformation of 0.0004535924 ton per pound, we calculate the captured emission in tons per day. When transforming barrels per day to tons per day, we multiplied the 400 kg CO_2 per barrel and the capture rate (85%) of CO_2 emitted to obtain the captured amount in tons per day. There is only one project reported in the unit cubic foot. The Bell Creek Project in Montana is a natural gas-enhanced oil recovery project. Based on the project summary, it plans to capture 900,000 metric ton of CO_2 per year so that it can be transformed to tons per day by dividing 900,000 metric ton of CO_2 per year by 365 days per year.

Based on the above two main indicators of CCS projects, I analyzed the spatial and temporal variations across all the projects. In addition, I analyzed the relationship between the project cost and capture and storage amount using linear regressions, both at a global level and for each different region. Finally, I summarized the factors affecting the success or failure of the CCS projects by summarizing the reasons for the failure of different projects documented in the CCUS database.

4 Results

4.1 Summary of CCS Projects in the Database

Most of the projects in the database are in the "planning" phase, followed by the "operational" phase, and then by "engineering design" and "plant design" (see Fig. 1). Of the 388 projects in total, 89 of them are planning-phase projects, 48 of them are operational-phase projects, 32 of them are engineering-design-phase projects, and 29 of them are plant-design-phase projects. Among the projects on other stages, the amount of projects that is on the phase of site characterization, permitting, developing infrastructure, capture ongoing, capture complete, injection ongoing, and injection complete are around 20, respectively. There is only less than or equal to five projects in site selection, feasibility study, memorandum of understanding,

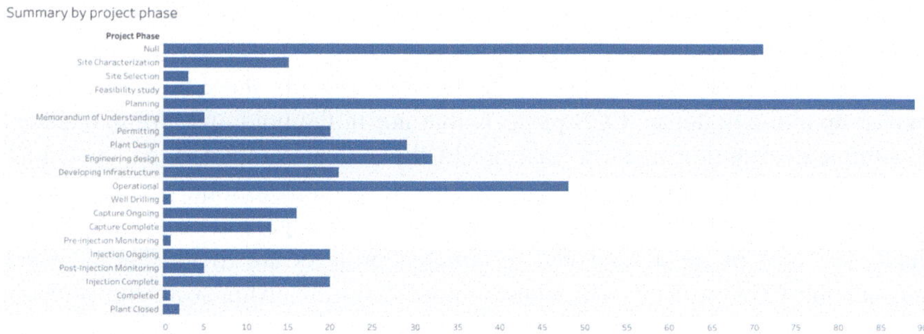

Fig. 1 Counts of projects by project phase

well drilling, pre-injection monitoring, post-injection monitoring, plant closed, and, especially, completed phase. Just as analysis of the chart above, only a small amount of projects entered the later phase of development. High percentage of CCS projects around the world is staying in the planning phase and different design phases such as plant design and engineering design. Seventeen percent is having missing values in the project phase.

4.2 Cost Distribution for CCS Projects

Cost is an important aspect of CCS projects. I first evaluated the cost distribution across different statuses of CCS plants.

The average cost of completed CCS projects is dramatically lower than the average cost of active, hold, potential, and terminated phase projects. Specifically, projects that are on hold and terminated have average cost around 1000 million, while the average cost of active-phase and potential-phase projects are around 700 million USD, displaying a significant difference (see Fig. 2, left panel). In particular, the average cost of completed projects is only below 50 million USD.

Judging from the distribution of cost (see Fig. 2, right panel), projects that are in active and potential phases include multiple outliers in cost. Specifically, there are 10 out of 65 projects with extremely high cost that are shown as outliers for those in the active phase and 5 out of 21 projects in potential phase categorized as outliers due to high cost. Some outlier projects even cost over 10 billion USD. As such, active and potential CCS projects are highly diverse in costs and are skewed toward the high end of costs. Without those outliers, it is obvious that active-phase, completed-phase, and potential-phase projects have cost that is far lower than 1 billion USD foremost. By contrast, projects that in hold and terminated stages mainly ranges from 0 to nearly 5 billion USD. These results indicate significant project-specific cost aspects, which raises very high requirements for funding and capital for some CCS projects. I further discuss this in Discussion.

I then examined the geographical distribution of CCS plants by their costs (see Fig. 3). CCS projects that are in active, completed, and potential stages distribute mainly in America, some European countries, north-east China, and Japan. Relatively high-cost projects only occur in America, Europe, and Australia. Low-cost projects have a high percentage among project in these three stages and gather comparatively dense. CCS projects that are in the hold and terminal stages have similar distribution to those that are in the active, completed, and potential stages. They are mainly concentrated in America (especially in southwest America), Europe, and east Australia. When looking into the average cost in USD of these projects, the cost of most hold and terminated projects are in the middle level among all the CCS projects which are around 6 billion. Almost all the projects in these stages cost at an average value. By contrast, projects that are in the active, completed, and potential stage have a huge gap between each other. A few individuals cost over around 30 billion, while others cost far less than ten million.

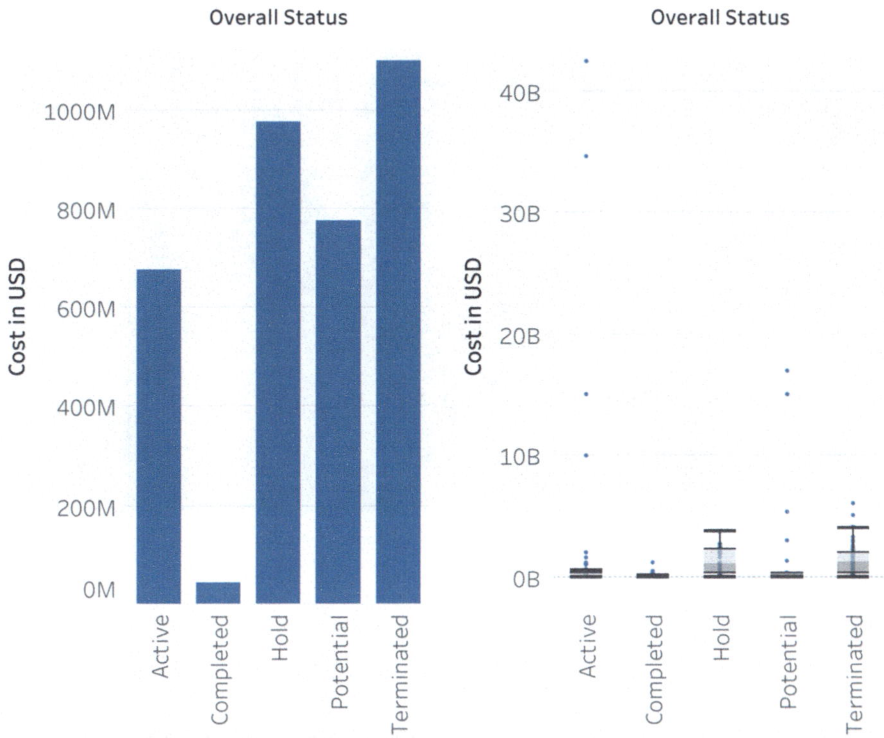

Fig. 2 Average costs of CCS projects by overall status (left) and the box plot for the distribution of project costs of CCS projects by overall status (right). In the box plot, the middle cover divergence indicates the median of the cost distribution, and the thin black lines represent the 25th and 75th percentile. The thick black lines indicate the 5th and 95th percentile. The dots indicate the cost values outside of the 5th and 95th percentile

Among active projects, the highest cost ones are the CO_2 Europipe in Europe and Gorgon project in Australia. The CO_2 Europipe project is a consortium that plans to initialize the infrastructure, enabling large-scale, European-wide transport and injection of CO_2 from zero-emission plants. It targets 2020 as the year for the start of the large-scale CCS in Europe. Due to its great application field and scope that ranges among Europe, it will definitely have an extremely high cost. The Gorgen project is one of the world's largest LNG projects and the largest single resource project in Australia's history. Located in Barrow Island—a Class A Nature Reserve—Gorgon comprises a three-train, 15.6 million ton per annum LNG facility and a domestic gas plant. In order to supply gas to Western Australia and meet the global demand for cleaner-burning fuel, there will be a huge expense on transportation over the ocean and make this project one of those with the highest cost.

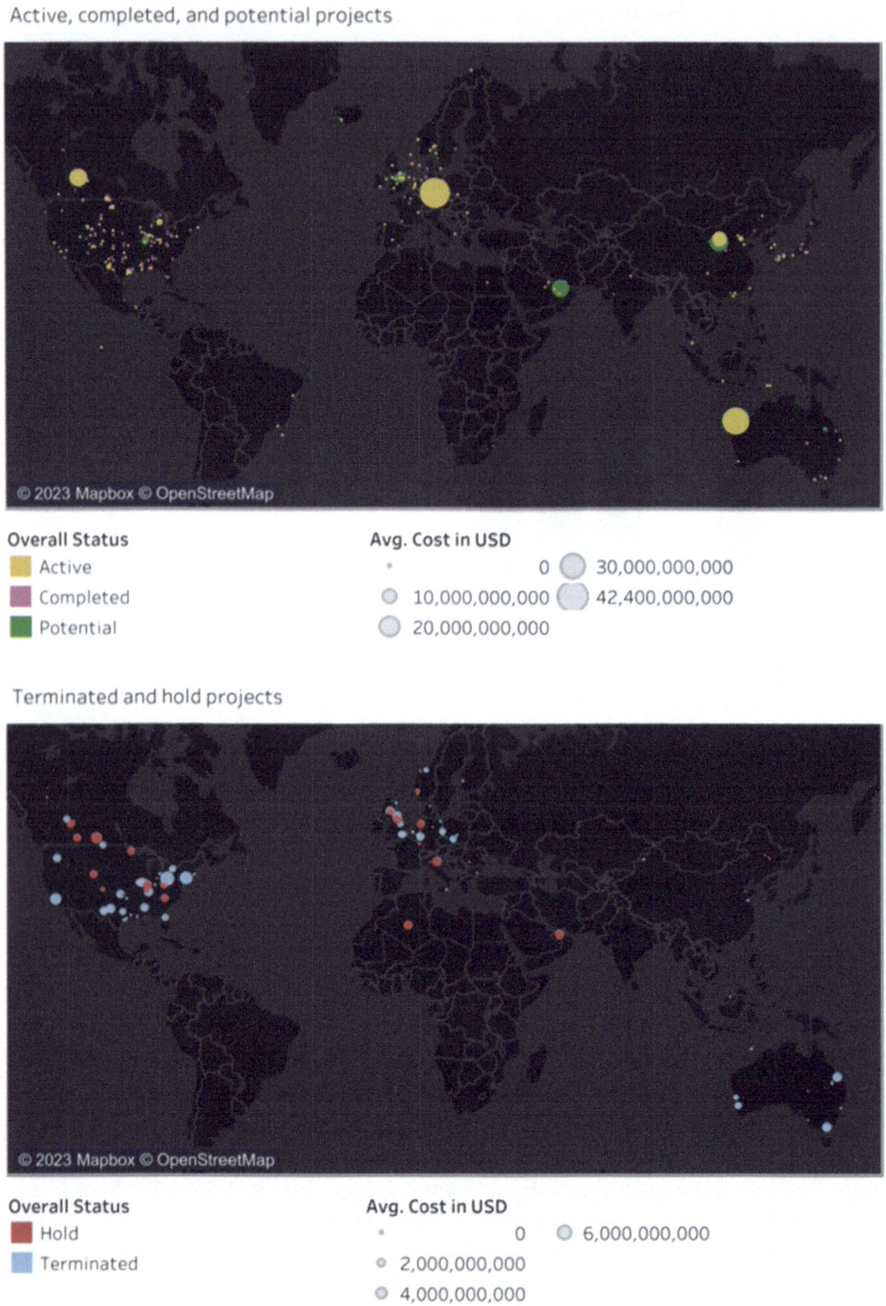

Fig. 3 Overall status and average cost of CCS projects from countries around the world

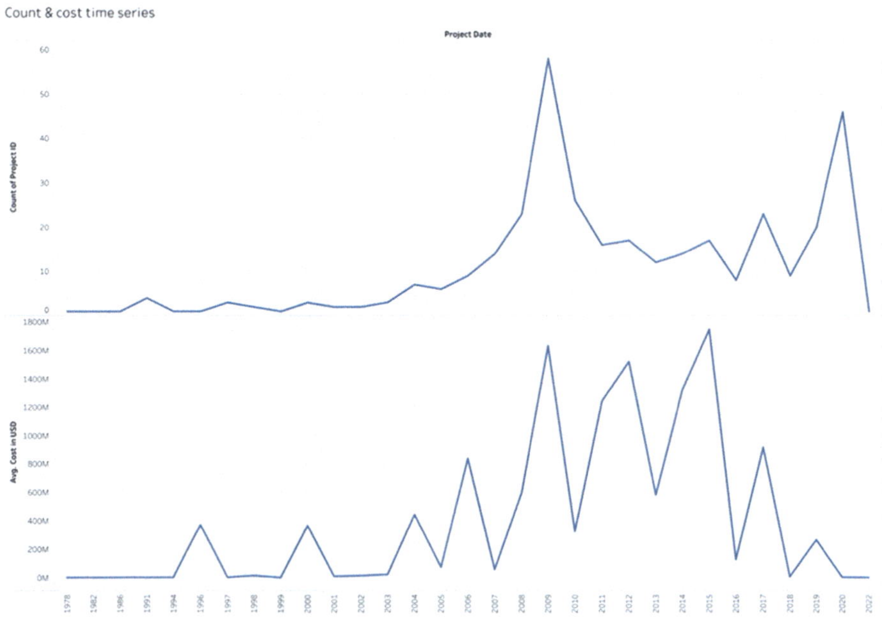

Fig. 4 Count of CCS projects by project date (top) and average cost of CCS projects by project date (bottom)

Finally, costs vary with time. The top panel of Fig. 4 shows the total number of CCS projects changing over time. The bottom panel of Fig. 4 shows the trend of cost with time.

There are two surges in the number of projects: 2009 and 2020. Before 2007, there were only less than ten or nearly no CCS projects. However, starting from 2007, the count of CCS projects shoot up to around 60 in 2 years and suddenly fell down to around 15 in the following 2 years. While fluctuating for several years, the count of CCS projects surges again to nearly 45 in 2020.

The cost of project has been increasing since 2004, but has been rated low in the recent 2 years. Fluctuations with large amplitude continuously occur in the average cost of the project.

4.3 *Captured Amount and Its Relationship with Cost*

The captured amount of CCS projects is another important dimension to evaluate the projects.

When evaluated by plant status, projects in the active phase appear to have the highest capture amount (see Fig. 5). Their average daily CCS amount is nearly 2300 tons per day—more than twice as much as those in the terminated phase, projects

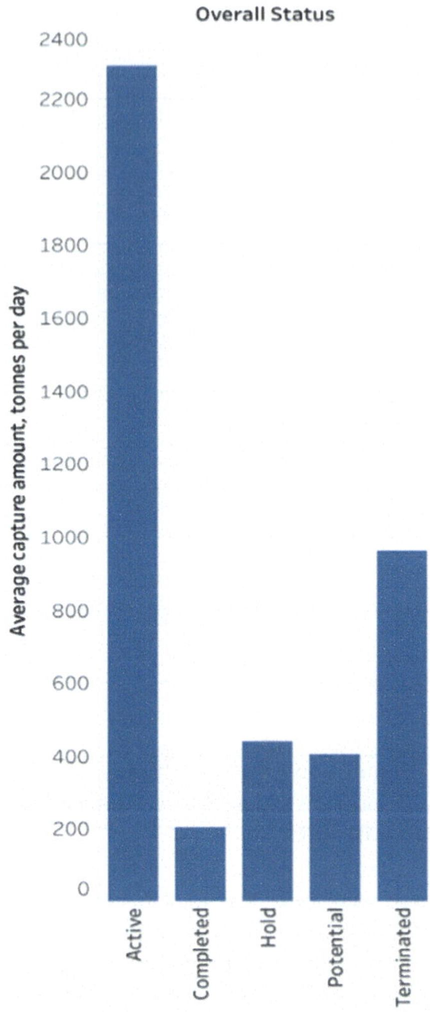

Fig. 5 Average daily CCS amount in tons per day by plant status

with the next highest capture amount. These terminated projects are supposed to capture around 1000 tons of CCS amount per day on average. The rest of the projects include those completed, hold, and potential and show far less captured amount lower than 500 tons per day. Among these projects, completed projects have the lowest capture amount for around 200 tons per day, indicating that the development and implementation of CCS technologies are still in the initial phase.

As a next step, we investigate the statistical relationship between cost and captured amount (see Fig. 6). As shown by the chart, the p-value is less than 0.0001, indicating high statistical significance between project cost and captured amount (see Table 1).

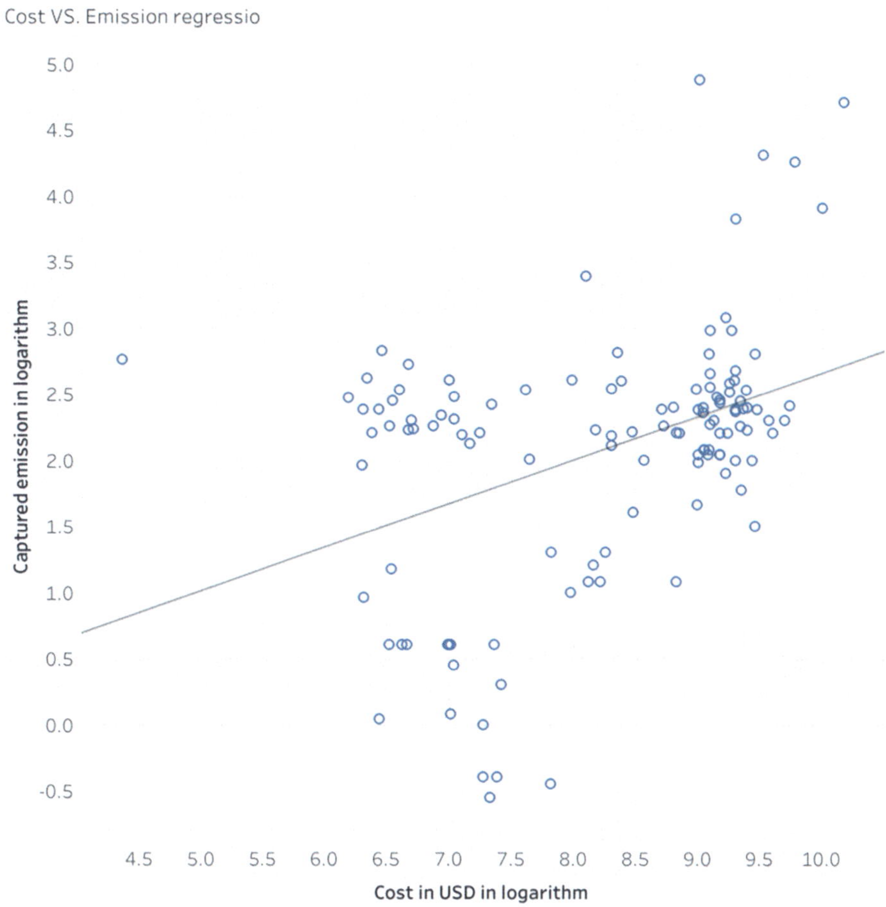

Fig. 6 Relationship between cost of projects and captured emissions for all CCS projects globally

Among the projects in the CCS inventory, when the project cost increases by 1%, the captured amount of the projects usually increases by 0.33% correspondingly. Not surprisingly, the higher the capital investment, the larger the scale of capture storage. However, our results further demonstrate the capital-intensive nature of CCS projects. That is, to further expand the scale of capture and storage by 1%, 3% more costs are involved. This is especially substantial for the already large-scale plants, because the percent cost increase means even higher cost in absolute levels when the base levels are high already.

In addition, we examine the same relationship for different countries and regions (see Fig. 7). Canada has the highest coefficient, which is 1.26534, which means when the cost of the project in Canada increases to 1%, the capture amount will increase to around 126.5% accordingly, indicating the highest return to investment. The United Kingdom has the second highest coefficient, i.e., 0.990184, which also

Table 1 Results of the statistical analysis between cost of project and captured amount

Region	Term	p-value	DF	Value	StdErr
Global	Log(cost)	<0.0001	120	0.329585	0.0687436
	Intercept	0.267597		−0.636204	0.571208
Australia	Log(cost)	0.17195	12	0.337502	0.232327
	Intercept	0.23021		−0.622845	1.88518
Canada	Log(cost)	0.63206	7	1.26534	0.962833
	Intercept	0.0044225		−8.99092	8.90926
China	Log(cost)	0.00702	1	0.27928	0.428175
	Intercept	0.0106521		0.482577	3.731
Continental Europe	Log(cost)	0.17195	32	0.430452	0.140543
	Intercept	0.23021		−1.51557	1.12807
United Kingdom	Log(cost)	0.63206	4	0.990184	0.194446
	Intercept	0.0044225		−6.6884	1.73967
United States	Log(cost)	0.00702	48	0.280473	0.105529
	Intercept	0.0106521		−0.299442	0.879852

demonstrates high return to the cost. Among other countries, the coefficient ranges between 0.28 and 0.43, thus indicating a 28–43% return for 1% increase in cost. As such, in all major regions worldwide, the increase in CCS costs appears to bring in much higher rise in capture amount. This corresponds to the capital-intensive nature of the projects. The scale of the CCS projects plays a major role in the outcomes of capture and storage.

4.4 Captured Amount and Its Relationship with Cost

Finally, I investigated the reasons for projects that have been on hold and terminated, based on the summary provided by the database and other sources. Commons reasons of unsuccessful implementation include the following:

Financial concerns.

CCS projects requires high capital cost upfront, and the construction costs can be high depending on the locations, such as the Belchatow CCS Project and E.ON Kingsnorth Ruhrgas UK Post-Combustion Project.

Insufficient economic benefits in the carbon trading market and uncertainty in demand, such as the Pioneer Project.

Uncertainty in the economic environment, such as the RWE nPower-Blyth Post-Combustion Project.

Difficulty in obtaining permits.

Multiple projects fail due to the permitting and legislation issues, such as the Porto Tolle and the American Electric Power-Mountaineer projects.

Unclear role of political support.

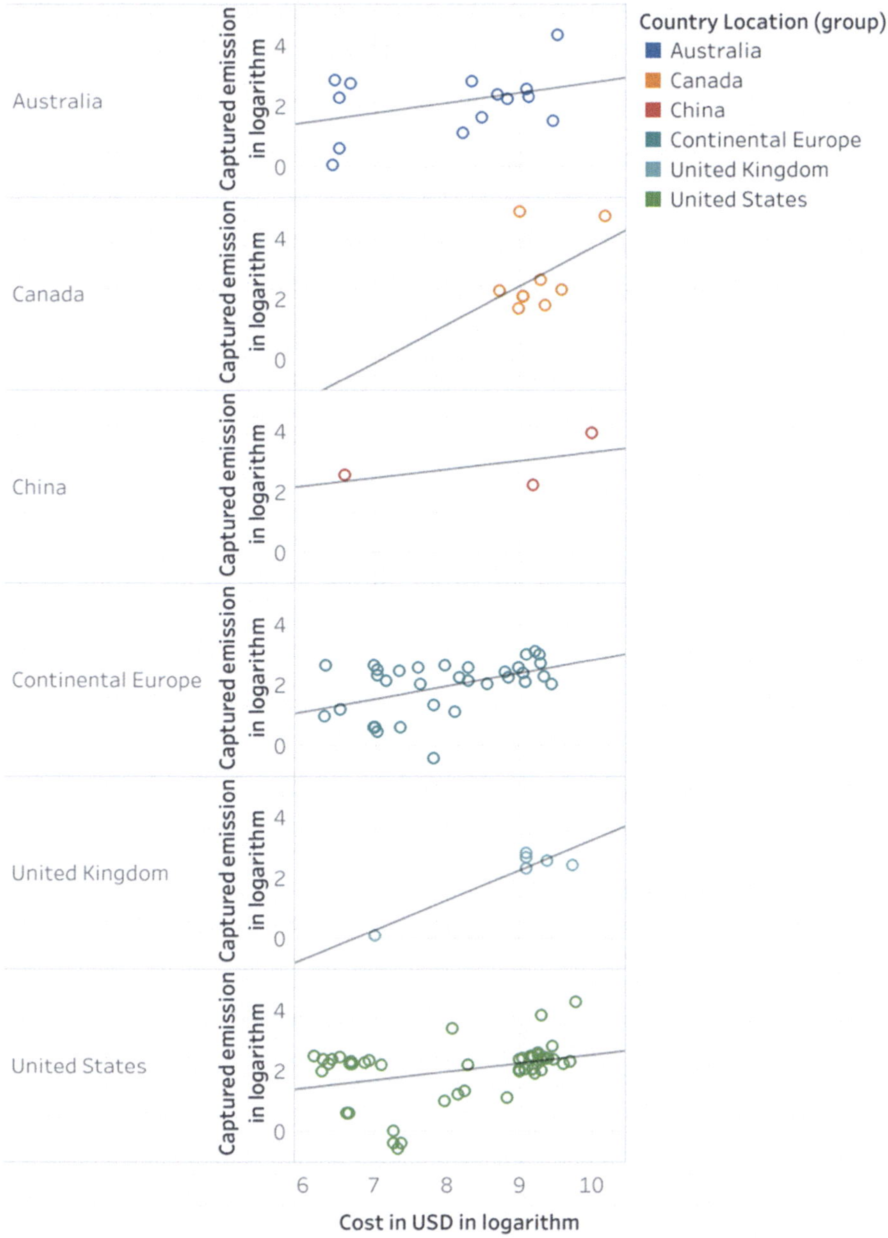

Fig. 7 Results of regressions by regions

Projects can fail due to lack of political support or uncertainty in climate policy, such as the DONG Energy-High-Efficiency PC Plant in Greifswald.

5 Discussion and Conclusion

In this paper, I used the global inventory of CCS projects to investigate the costs and capture amounts of CCS projects over time and geographical locations. I found that cost is a key factor for determining the progress or execution of the CCS projects. Among all the CCS projects, the terminated projects are usually with extremely high costs. The captured amount of emission is usually larger with higher cost involved. For projects with large scale in capture and storage, increase of cost has a larger increasing effect on the capture and storage capacity. This demonstrates the capital-intensive nature of the CCS industry. As a result, smaller and less capital invested CCS projects may be difficult to succeed and advance.

As the climate stress becomes increasingly urgent, understanding and use of CCS is in high demand, yet still only studied and implemented with limited efforts. It is necessary to understand the factors that either facilitate or hinder the development of CCS projects and learn the comprehensive picture of this technology.

One of the key factors this paper shows on the success of CCS projects is the cost and financing issue. This has been one of the main reasons that causes CCS projects to be terminated. It has been found that significant reduction on the cost of CCS projects requires sustained development and substantial extent of commercial deployment. However, these requirements should be based on a significant market that has not been built yet. Also, the longer the estimated cost of CCS projects exceed the market price, the longer uncertainty of CCS projects' social acceptance lasts, which further lead to the lack of funding from the government to reduce the gap of costs [8]. Fortunately, in order to reach emission mitigation, reward mechanisms have been considered within the UNFCCC framework. This can cope with and alleviate CCS projects' barrier of costs [9].

This research can provide people a preliminary understanding of CCS, realizing its significance for mitigating environmental issue and knowing the factor of the success or failure of CCS projects in order to solve them and apply CCS in a wider range.

References

1. Boot-Handford, M.E., Abanades, J.C., Anthony, E.J., Blunt, M.J., Brandani, S., Mac Dowell, N., Fernández, J.R., Ferrari, M.-C., Gross, R., Hallett, J.P., Haszeldine, R.S., Heptonstall, P., Lyngfelt, A., Makuch, Z., Mangano, E., Porter, R.T.J., Pourkashanian, M., Rochelle, G.T., Shah, N., Yao, J.G.: Carbon capture and storage update. Energy Environ. Sci. **7**(1), 130–189 (2014). https://doi.org/10.1039/c3ee42350f

2. Ma, J., Li, L., Wang, H., Du, Y., Ma, J., Zhang, X., Wang, Z.: Carbon capture and storage: history and the road ahead. Engineering. **14**, 33 (2022). https://doi.org/10.1016/j.eng.2021.11.024
3. Shu, D.Y., Deutz, S., Winter, B.A., Baumgärtner, N., Leenders, L., Bardow, A.: The role of carbon capture and storage to achieve net-zero energy systems: trade-offs between economics and the environment. Renew. Sust. Energ. Rev. **178**, 113246 (2023). https://doi.org/10.1016/j.rser.2023.113246
4. Lau, H.C., Ramakrishna, S., Zhang, K., Radhamani, A.V.: The role of carbon capture and storage in the energy transition. Energy Fuel. **35**(9), 7364–7386 (2021). https://doi.org/10.1021/acs.energyfuels.1c00032
5. IEA.: CCUS Projects Database – Data product. [online]. Available at: https://www.iea.org/data-and-statistics/data-product/ccus-projects-database (2023)
6. Google.: Google Finance. [online] www.google.com. Available at: https://www.google.com/finance/ (2023)
7. US EPA, O.: Greenhouse Gases Equivalencies Calculator – Calculations and References. [online]. Available at: https://www.epa.gov/energy/greenhouse-gases-equivalencies-calculator-calculations-and-references#:~:text=Home%20electricity%20use&text=The%20national%20average%20carbon%20dioxide (2015)
8. Karayannis, V., Charalampides, G., Lakioti, E.: Socio-economic aspects of CCS technologies. Proc. Econo. Financ. **14**, 295–302 (2014). https://doi.org/10.1016/s2212-5671(14)00716-3
9. Zakkour, P., Dixon, T., Cook, G.: Financing early opportunity CCS projects in emerging economies through the carbon market: mitigation potential and costs. Energy Procedia. **4**, 5692–5699 (2011). https://doi.org/10.1016/j.egypro.2011.02.563

Energy-Food Commodity Price Movement and Equation in Kyoto (2006) Perspective: Factor Analysis and Nonlinear Framework

Aynur Pala and Bahia Bouchafaa

1 Introduction

Energy and food price are the most important production factors and investment tools. Indeed, with economic and social development, food production increasingly requires specialized economic agents responsible for producing, storing, preserving, transforming, transporting and marketing products in order to provide healthy, high-quality, easy-to-consume and store, sufficient and regular food to the largest possible population. Thus, the food supply of developed countries mobilizes a significant part of the means of production and would represent 20–25% of the total energy consumption. Then, analyzing the trend and relation equation of energy and food price is important for producers, researchers, and policymakers to manage risk because its price movements determine the input cost of the world economy. However, the Kyoto conference marked awareness of the real risk represented by the emission of greenhouse gases linked in particular to the consumption of fossil energy and its consequences on global warming [1]. On the other hand, energy and food prices are affected by a lot of endogeneous and exogeneous factors. The rising petroleum price and environmental risk lead to alternative sustainable energy source as biofuel. After the Kyoto Protocol (2006), the rising bioenergy production and drought cause high agricultural commodity prices. There was a rise in crude oil and agricultural commodity prices in 2008.

A. Pala (✉)
Istanbul Okan University, Istanbul, Türkiye
e-mail: aynur.pala@okan.edu.tr

B. Bouchafaa
National Polytechnic School of Algiers, El Djazaïr, Algeria

L. Chen (ed.), *Trends in Clean Energy Research*, Green Energy and Technology,
https://doi.org/10.1007/978-3-031-67987-2_23

Before proposing our methodology, we deemed it necessary to analyze the situation of the movements of energy and food commodity prices, with the aim of arguing and justifying our work. That is what we present in the next session.

2 Analysis of the Situation

Figure 1 shows the evolution of the energy and food commodity price indices over the period 1991–2020.

We focus on the Kyoto Protocol (2006) treatment to examine the relationship between energy and food commodity price. The food price crisis started after Kyoto (2006) due to climate change and especially beginning nonfood use of agricultural production based on the hot topic trend, "decreasing carbon emission."

Some studies have examined the relationships between energy and food prices. Yu et al. [2] examined the causal link between crude oil and vegetable oil prices. They found a relatively low effect of oil on vegetable prices. Tyner [3] analyzed the relationship between energy and agricultural product prices. They found a low correlation pre-2005 and a strong correlation in mid-2008. Chen et al. [4] show unidirectional relationship running from oil prices on grain prices. Pala [5] examined the relationship between food and energy prices by using the Granger causality in the structural break perspective. They found that the cointegration regression coefficient is negative in pre-2008 crisis, but adversely positive in post-2008. These results represent that the relationship between crude oil and food price equations has changed. Pala [6] examined the energy and food price relationship by using the linear Granger causality method in the Kyoto (2006) perspective and found that during the post-Kyoto period, there is unidirectional running from food price to energy price which is the exact opposite in the pre-Kyoto period. Cheng

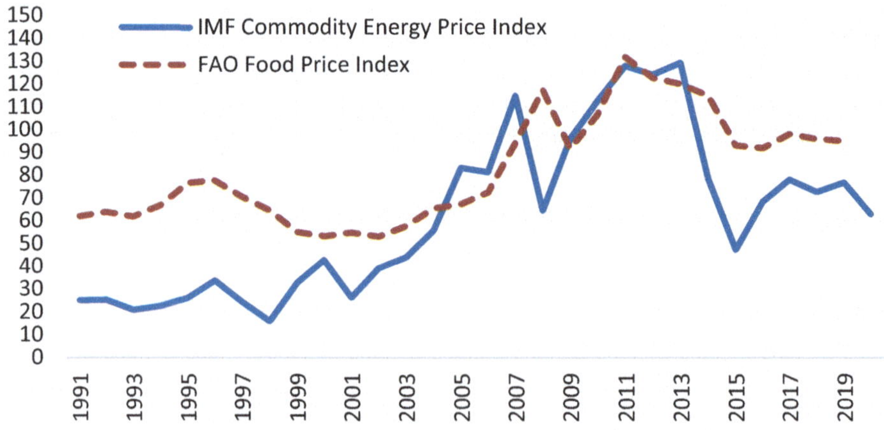

Fig. 1 Energy and food commodity price movement, 1991–2020

and Cao [7] examined dynamic relationship between global crude oil price and food price indices using TVAR and TVECM for the period from January 1990 to June 2017 within a nonlinear framework. They found that there is a nonlinear causal relationship between global crude oil and food price indices.

This aroused our curiosity to try to verify more closely the real relationship between energy and food prices as well as the direction of causality. As our concern is twofold, i.e., to verify the causality and its direction, we first studied the time series of the said prices and used statistical methods to answer our questions. In this paper, we try to determine and explain the impact of the 2006 Kyoto Protocol on nonlinear causality between energy and food commodity prices, using statistical methods in order to distinguish between two factors, i.e., pre-Kyoto and post-Kyoto Protocol (2006).

3 Data and Methodology

As explained above, and in order to verify the impact of the 2006 Kyoto Protocol on the behavior of causality between energy and food prices, if there is causality, we studied their series and their co-variabilities. To do this, we used several statistical methods that we deemed necessary to verify the behavior of the movement of these prices.

First, we analyzed the relation between energy and food commodity prices by using proxy variables for each commodity sub-block. The factor indices are calculated by the factor analysis method on FAO food price contents, e.g., meat, dairy, cereals, vegetables oils, and sugar, and IMF Energy Price Index contents, e.g., fuel oil (WTI), natural gas (Henry Hub), and coal. These variables are summarized in Table 1.

Then, we analyzed the time series properties and relationship among commodity trio sub-block prices. Moreover, we defined the best proxy variables from all contents of the trios by using the factor analysis, which is a dimension reduction method. International commodity price indices and all their contents used in the model are shown in Table 2. The data covered the period from 1991 to 2020 due to some limited data series.

We used new factor index proxy variables instead of each commodity duo by using the factor analysis method on all contents of international commodity price indices: FAO Food Price Index and IMF Commodity Energy Price Index. The factor

Table 1 Energy and food commodities variables

	Contents of international commodity price Indices
Energy	IMF commodity energy Price index: Oil (WTI, Brent crude oil), natural gas (Henry hub), and coal price
Food and beverage	FAO food Price index: Meat, dairy cereals, oils, and sugar price

Source: IMF

Table 2 Proxy variable definition

Proxy variable names and codes	Calculation criteria and source
Energy Price Factor Index, EPFI	Calculated using contents of IMF commodity energy Price index with factor analysis
Food Price Factor Index, FPFI	Calculated using contents of FAO food Price index by factor analysis

analysis method is used to identify a reduced number of variables and calculate factor indices. In social science, factors usually explain 50–60%.

Brock, Dechert, and Scheinkman (BDS) [8] and Tsay [9] tests were used to test non-linearity. Enders and Granger [10] and Kapetanios, Shin, and Shell (KSS) [1] nonlinear unit root tests were also used. The optimal lag length for cointegration is determined by Akaike's information criterion (AIC) [12], Schwarz's Bayesian information criterion (SIC) [13], Hannan-Quinn criterion (HQC) [14], final prediction error (FPE) [15], and Bayesian information criterion (BIC) [16]. Kapetanios, Shin, and Snell [17] nonlinear cointegration test and Diks-Panchenko [18] nonlinear causality tests were also used.

As a support for our study, we used the content of international energy and food price indices as price variables for meat, dairy products, cereals, oils and sugar, crude oil, natural gas, and coal.

4 Results

After carrying out the methods and techniques that we proposed, we obtained the following results.

The food price factor index as a proxy was generated from FAO meat, dairy, cereals, oils, and sugar price indices by using the factor analysis. The proxy variables are shown in Table 2. The energy price index is calculated by using the factor analysis on IMF crude oil, natural gas, and coal prices. We presented cumulative explaining ratios by 89% for the energy price factor index. The total factor explanation cumulative proportion increased to 97% with the food price factor index.

Enders and Granger [10] nonlinear unit root tests examine the null hypothesis of a "unit root" against the alternative of threshold adjustment. The test results, presented in Table 3, show that all factor index variables are stationary in level. It was supported by Kapetanios, Shin, and Shell (KSS) [11] nonlinear unit root test results.

The KSS nonlinear cointegration [17] test is presented in Table 4 with for raw (case = 1), demeaned (case = 2), and detrended (case = 3) models. The results show that there is a cointegration relation between energy and food price factor indices.

The optimal lag length for cointegration determined by Akaike information criterion (AIC) [12], Schwarz's Bayesian information criteria (SIC) [13], Hannan-

Table 3 Enders and Granger [10] MTAR and KSS (2003) nonlinear unit root tests

Variables	Enders and Granger [10] MTAR	KSS (2003)
EPFI	1.810***	−2.741***
FPFI	1.687***	−2.969**

*1%, **5%, ***10% critical values −2.59, −1.90, −1.57, Enders and Granger [10] p1 = p2 = 0 unit root
1%, **5%, ***10% critical values −3.47, −2.93, −2.66 KSS (2003)

Table 4 Kapetanios et al. [17] nonlinear cointegration test results

Dependent variables	Independent variables	Case = 1 raw	Case = 2 demeaned	Case = 3 detrended
EPFI	FPFI	−3.122	−3.243	−3.860***
FPFI	EPFI	−3.181***	−3.230	−3.757***
	1%	−4.14	−4.38	−4.70
Critical values	5%	−3.47	−3.77	−4.07
	10%	−3.15	−3.36	−3.75

*, ** and *** Indicate significance at the 1, 5, and 10% levels, respectively. The critical values for the test statistic were obtained from Kapetanios et al. [17]

Table 5 Diks and Panchenko [18] nonlinear causality results in full and subsamples

Full sample								
Ho hypothesis	1	2	3	4	5	6	7	8
FPFI does not cause EPFI	0.691	0.402	0.105	0.308	0.222	0.274	0.247	0.362
EPFI does not cause FPFI	0.011[a]	0.009[a]	0.020[a]	0.015[a]	0.041[a]	0.073	0.069	0.067
Subsamples								
Pre-2006: Ho hypothesis	1	2	3	4	5	6	7	8
FPFI does not cause EPFI	0.952	0.813	0.747	0.432	0.076	0.062	0.045[a]	0.045[a]
EPFI does not cause FPFI	0.201	0.237	0.277	0.035[a]	0.052	–	–	–
Post-2006: Ho hypothesis	1	2	3	4	5	6	7	8
FPFI does not cause EPFI	0.788	0.362	0.067	0.209	0.159	–	–	–
EPFI does not cause FPFI	0.008[a]	0.040[a]	0.084	−0	−0	–	–	–

[a]indicate the rejection of the null hypothesis of the absence of causality at the 5% level

Quinn criterion (HQIC) [14], and final prediction error criteria (FPE) [15] is 3. We used Diks-Panchenko [18] test to analyze nonlinear causality. The Diks-Panchenko [18] nonlinear causality test results are presented in Table 5.

The nonlinear causality in full-sample results obtained from Diks and Panchenko [18] indicate that the null hypothesis of "energy price does not cause food price" can be rejected. This implies the presence of a unidirectional causal relationship from energy to food price. In the pre-Kyoto (2006) subsample, nonlinear causality test results show that there is no nonlinear causality relation between them. But in the post-Kyoto (2006) subsample, there is a nonlinear causal relationship running from the energy price to food price factor index.

5 Concluding Remarks

In order to verify the existence and determine the direction of causality between the prices of energy and food commodity, we considered two distinct periods: pre-Kyoto and post-Kyoto 2006. Indeed, after noting a change of the behavior of the movement of these prices between these two periods, we considered it necessary to study their co-variability according to the two factors previously mentioned.

We studied the relationship between energy and food prices generating two proxy factor price index variables instead of energy and food prices. The nonlinear causality in full-sample results obtained from Diks and Panchenko [18] indicate that the null hypothesis of "energy price does not cause food price" can be rejected. This implies the presence of a unidirectional causal relationship from energy to food price.

The nonlinear causality test in full-sample results show that there is a unidirectional causal relationship running from energy to food price. In the pre-Kyoto (2006) subsample, nonlinear causality test results show that there is no nonlinear causality link among them. In the post-Kyoto (2006) subsample, there is a nonlinear causal relationship running from the energy price to food price factor index. The analysis that we carried out on the time series of prices of energy and raw food materials allowed us to conclude about the presence and linear direction of causality between them. Also, it was possible for us to determine the impact of the 2006 Kyoto Protocol on the behavior of these price movements. Our findings are similar to Tyner [3]. They found a low correlation pre-2005 and a strong correlation in mid-2008 between energy and food prices. The findings are also supported by the study of Cheng and Cao [7].

Finally, this study allowed us to verify and prove the impact of the 2006 Kyoto Protocol on movements in the prices of energy and raw food materials as well as the direction of causality between them. This approach can provide further insights into the dynamics of commodity prices and their interdependencies. Let us also note that our study can be the gateway to the development of a model describing the variation in the prices of food commodity as a function of those of energy, taking into account the policies adopted for the preservation of the environment and the adoption sustainable and renewable energy. We recommend for future studies about between energy and food price to extent of the three main commodity sub-blocks: energy, food, and industrial raw materials.

References

1. Monot C.: Quel est le coût énergétique de notre alimentation? Club Energie, Prospectives et Débats. Groupe 1. Claire Monet – Cirad amis Ecopol à la demande du CNRS, p. 2 (2000)
2. Yu, T-H., Bessler, D.A., Fuller, S.: Cointegration and causality analysis of world vegetable oil and crude oil prices. In: American Agricultural Economics Association Annual Meeting, Long Beach, CA, July 23–26 (2006)

3. Tyner, W.E.: The integration of energy and agricultural markets. Agric. Econ. **41**, 193–201 (2010)
4. Chen, S.T., Kuo, H.I., Chen, C.C.: Modeling the relationship between the oil price and global food prices. Appl. Energy. **87**, 2517–2525 (2010)
5. Pala, A.: Structural breaks, cointegration, and causality by VECM analysis of crude oil and food price. Int. J. Energy Econ. Policy. **3**(3), 238–246 (2013)
6. Pala, A.: Have Global Warming and Measures Taken Changed the Relationship between Food and Oil Prices? (In Turkish) (Küresel Isınma ve Alınan Önlemler Gıda Petrol Fiyatları İlişkisini Değiştirdi mi?)", Okan Universitesi Press (2018)
7. Cheng, S., Cao, Y.: On the relation between global food and crude oil prices: an empirical investigation in a nonlinear framework. Energy Econ. **81**, 422–432 (2019)
8. Brock, W.A., Dechert, W.D., Scheinkman, J.A., LeBaron, B.: A test for independence based on the correlation dimension. Econ. Rev. **15**, 197–235 (1996)
9. Tsay, R.S.: Testing and modeling multivariate threshold models. J. Am. Stat. Assoc. **93**(443), 1188–1202 (1998)
10. Enders, W., Granger, C.W.J.: Unit-root tests and asymmetric adjustment with an example using the term structure of interest rates. J. Bus. Econ. Stat. **16**, 304–311 (1998)
11. Kapetanios, G., Shin, Y., Snell, A.: Testing for a unit root in the nonlinear STAR framework. J. Econ. **112**(2), 359–379 (2003)
12. Akaike, H.: Information theory and an extension of the maximum likelihood principle. In: Petror, B.N., Csaki, F. (eds.) Second International Symposium on Inference Theory, pp. 267–281. Akademia Kiado, Budapest (1973)
13. Shwartz, G.: Estimating the dimension of a model. Ann. Stat. **6**, 461–464 (1978)
14. Hannan, E.J., Quinn, B.G.: The determination of the order of an autoregression. JRSS. **B41**, 190–195 (1979)
15. Akaike, H.: Fitting autoregressive models for prediction. Ann. Inst. Stat. Math. **21**, 243–247 (1969)
16. Akaike, H.: A Bayesian extension of the minimum AIC procedure of autoregressive model fitting. Biometrika. **66**, 237–242 (1979)
17. Kapetanios, G., Shin, Y., Snell, A.: Testing for cointegration in nonlinear smooth transition error correction models. Economet. Theor. **22**(2), 279–303 (2006)
18. Diks, C., Panchenko, V.: A new statistic and practical guidelines for nonparametric granger causality testing. J. Econ. Dyn. Control. **30**, 1647–1669 (2006)

Sustainable Construction Materials from Industrial By-products

Sawsan Dagher, Khalefah Alabdouli, Ahmead AlHosani, Yousif Hnan, Khalid Aljneibi, Mohamed Binsafwan, Boshra Akhozheya, and Maisa El Gamal

1 Introduction

In the realm of sustainable construction, waste materials offer a promising solution to environmental concerns while meeting infrastructure demands. Among these, waste sustainable concrete stands out, reducing carbon emissions, conserving resources, and minimizing waste. This concrete incorporates industrial by-products and recycled materials, reducing reliance on traditional materials. Notably, it repurposes waste such as fume treatment plant (FTP) dust and mixture dust of direct reduction iron of (DRI) with lime as a supplementary cementitious material [1]. FTP dust has the most conspicuous environmental consequence by steel manufacturing, which is an airborne pollution discharged from factory smokestacks, characterized by swirling clouds of gray and black emissions. These emissions contain a variety of pollutants, spanning from solid metallic particles to volatile organic substances [2]. Handling dust of DRI and lime poses hazards including inhalation risks, skin and eye irritation, fire and explosion potential, chemical reactivity, and storage-related dangers [3]. Additionally, recycled aggregates further decrease environmental impact by replacing virgin materials. Aligning with circular economy principles, waste sustainable concrete diverts materials from landfills, contributing to resource conservation. Moreover, it offers tangible benefits beyond environmental stewardship, including improved durability and reduced life cycle

S. Dagher (✉) · K. Alabdouli · A. AlHosani · Y. Hnan · K. Aljneibi · M. Binsafwan
Department of Electromechanical Engineering, Abu Dhabi Polytechnic, Abu Dhabi, UAE
e-mail: sawsan.dagher@adpoly.ac.ae

B. Akhozheya
Department of Architecture, Texas A & M University, College Station, TX, USA

M. El Gamal
College of Natural and Health Sciences, Zayed University, Abu Dhabi, UAE

L. Chen (ed.), *Trends in Clean Energy Research*, Green Energy and Technology,
https://doi.org/10.1007/978-3-031-67987-2_24

costs. Its potential for achieving green building certifications further incentivizes its adoption. Waste sustainable concrete presents an opportunity to balance economic growth with environmental responsibility in construction. By utilizing waste materials, we redefine concrete as a sustainable building material, preserving resources for future generations. This introduction lays the groundwork for exploring waste sustainable concrete's transformative potential in creating a more sustainable and resilient built environment [4]. In 2021, the global production of Portland cement predicted a modest rise, reaching an estimated 4.4 billion tons, up from 4.2 billion tons in 2020 [5].

However, despite this growth, expansion remains constrained due to various factors including rising service and material expenses, shortages in production and labor, ongoing disruptions in supply chains, and challenges in shipping and logistics. Sulfur concrete offers promise as an alternative construction material, showcasing unique properties beneficial for various applications. While progress has been made, challenges such as durability and environmental impact necessitate further research and innovation to fully exploit its potential in sustainable construction and infrastructure development [6]. Sustainable concrete exhibits remarkable properties including high chemical resistance, low permeability, efficient thermal conductivity, and rapid setting, making it well-suited for applications in corrosive environments such as chemical plants and marine structures. These properties ensure durability and longevity under harsh conditions, enhancing the material's appeal for various construction projects. Nonetheless, challenges such as susceptibility to temperature fluctuations and potential odor concerns during curing processes require careful consideration and mitigation strategies to optimize the performance and usability of sustainable concrete in diverse settings.

Concrete demonstrates versatile applications across hazardous material containment, paving, bridge decks, and industrial flooring, owing to its exceptional durability and resistance properties. Moreover, ongoing research endeavors are actively exploring its potential contribution to sustainable construction practices and infrastructure rehabilitation efforts. As a result, concrete emerges as a promising material offering both practical solutions for specialized construction needs and potential advancements in environmentally conscious building practices [7].

This study innovatively integrates sulfur, fine aggregates, and industrial waste into concrete, enhancing its durability and sustainability. By repurposing materials and minimizing environmental impact, it offers a greener alternative to traditional concrete. Rigorous testing confirms superior performance and lower chloride/sulfate levels, ensuring long-lasting structures resilient to environmental factors. This innovation paves the way for eco-conscious construction practices, promising a more sustainable future for infrastructure development.

2 Methodology

2.1 Materials and Equipment

Figure 1 shows that the materials employed in the production of sustainable concrete include sulfur (S) as a key binder, fine aggregate such as sand (FA) for enhancing structural integrity [8], fume treatment plant dust (FTP), and dust mixture of direct reduction of iron and lime (DRI and L) in powdered form to contribute to the concrete's strength and durability [9, 10]. Additionally, molds, typically sized at $5 \times 5 \times 5$ cm, are utilized for shaping the concrete into desired forms, ensuring precision and uniformity in construction projects., while heaters play a crucial role in the curing process, ensuring proper setting and hardening. Finally, blinder tools aid in the mixing and blending of the materials, facilitating the creation of high-quality sustainable concrete structures.

2.2 Preparation of the Concrete

The procedure for making sustainable concrete begins by heating sulfur (S) until it reaches a temperature of 120 °C [11] while simultaneously heating fine aggregate (FA), fume treatment plant dust (FTP), and dust mixture of direct reduction of iron and lime (DRI and L) on another plate. Once the materials are adequately heated, they are mixed together according to the specified sample compositions. For Sample 1, 350 g of sulfur, 330 g of fine aggregate, and 320 g of DRI and L are combined, ensuring thorough blending [12]. For Sample 2, the same amounts of sulfur and fine aggregate are used, with 320 g of FTP incorporated instead of DRI and L as shown in Table 1. After thorough mixing, the resulting mixture is poured into molds, shaped, and allowed to cool and set [13]. Once the concrete samples have solidified, they are carefully removed from the molds and subjected to stress testing to evaluate their mechanical properties. This procedure involves applying controlled force to the samples and measuring their response to determine factors such as compressive strength and durability [14] and also studying on the expansion of concrete under attack of sulfate and sulfate-chloride ions [15].

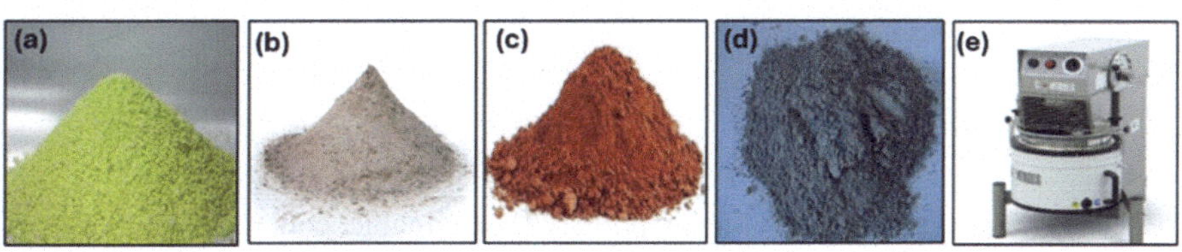

Fig. 1 Materials: (**a**) sulfur, (**b**) fine aggregate, (**c**) fume treatment plant dust, (**d**) dust of direct reduction of iron and lime, (**e**) laboratory planetary mixer with a heater

Table 1 Quantities of raw materials

Samples	S (g)	FA (g)	FTP (g)	DRI and L (g)
1	350	330	0	320
2	350	330	320	0

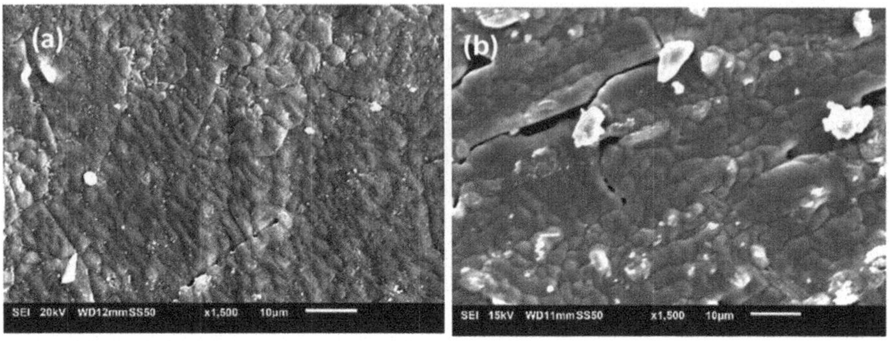

Fig. 2 Cross-sectional SEM analysis of sulfur concrete samples: (**a**) FTP and (**b**) DRI and L

3 Results and Discussion

3.1 Morphology and Structure

Figure 2 showcases a detailed cross-sectional SEM analysis of the sulfur concrete samples produced in this study. Images (a) and (b) correspond to the FTP and DRI and L samples, respectively. These micrographs vividly demonstrate the structural characteristics of the sulfur concrete, revealing how sulfur effectively binds, coats, and permeates between the aggregates. This process results in the formation of homogeneous sulfur concrete constructs, where the sulfur matrix intimately intertwines with the aggregate particles. Notably, the morphology of these samples closely mirrors that of conventional concrete, indicating the successful integration of sulfur without compromising structural integrity. Furthermore, the images highlight the smoothness and hardness of the sample surfaces, underscoring the robustness and quality of the sulfur concrete produced. Overall, this visual analysis provides valuable insights into the microstructural properties of sulfur concrete, validating its potential as a viable and effective construction material.

3.2 Mechanical Strength

The compressive strength of the as-prepared sustainable concrete samples was evaluated using a universal testing machine (UTM), as shown in Fig. 3. Table 2 displays the compression test results for the FTP and DRI and L samples. Compared to regular concrete, both samples exhibited superior performance, withstanding

Fig. 3 Compression testing machine

Table 2 Results of the compressive strength test

Samples	Density [kg/m^3]	Max. load [KN]	Compressive strength [MPa]
FTP	4402	46.220	19.3
DRI and L	4713	53.421	21.4

higher maximum loads due to the materials used in their composition. Sulfur played a key role in enhancing strength, resulting in increased durability and resistance to high loads [16].

The FTP sample demonstrated a compressive strength of 19.3 MPa, with a maximum load of 46,220 N, surpassing that of regular concrete. Studies have shown that FTP can improve both early-age and long-term strength by promoting pozzolanic reactions and enhancing durability [17]. On the other hand, the DRI and L sample exhibited even better results, with a compressive strength of 21.4 MPa and a maximum load of 53,421 N. This superior performance can be attributed to the presence of lime and metallic components in DRI dust, which interact with the cementitious matrix during hydration, resulting in a more durable and load-bearing concrete mixture [18, 19].

3.3 Chemical Analysis

The chemical composition of the DRI and FTP is provided in Table 3 [20, 21]. The chloride-sulfate test for the prepared concrete samples involves two distinct procedures for analyzing acid-soluble chloride and sulfate content, as depicted in Fig. 4.

For acid-soluble chloride determination, a 5 g concrete sample is dissolved in 100 mL of distilled water, followed by the addition of 10 mL concentrated nitric acid. The mixture is then boiled for 5 min, cooled, and filtered. Subsequently, 10 mL of 0.1 normal silver nitrate and 5 mL of ferric alum indicator are added

Table 3 Chemical composition of wastes

DRI [20]		FTP [21]	
Chemical composition	Percentage (%)	Chemical composition	Percentage (%)
Fe	67–69	Fe_2O_3	42.8
SiO_2	0.9–1.0	Zn O	34.4
Al_2O_2	0.2–0.3	CaO	3.85
CaO	1.1–1.2	MnO	2.90
MgO	0.80	SiO_2	2.81
S	0.015	PbO	1.92
PPb, Zn, Cu, Sn, Cr, As	0.0061.5	CuO, K_2O, Fe_2O_3	3.23

Fig. 4 Acid-soluble chloride and sulfate content

before titration against 0.1 normal ammonium thiocyanate until a red color appears. Calculations are then performed to determine the chloride content [22].

On the other hand, for acid-soluble sulfate analysis, a 5 g sample is mixed with 100 mL distilled water and 10 mL concentrated hydrochloric acid. After boiling for 5 min, the mixture is cooled and filtered. The filtrate is treated with 10 mL of 10% barium chloride solution, boiled again, and kept under boiling for 30 min. After cooling for 24 h, the residue is filtered through a fine-grade filter paper. The washed residue is then folded in the filter paper, ignited at 850 °C for 20 min, cooled in a desiccator, and weighed in a pre-weighted crucible. Subsequent calculations are conducted to determine sulfate content [23].

These methods provide valuable insights into the chloride and sulfate levels present in concrete, aiding in quality control and assessment of potential environmental impacts.

The calculations for sulfate and chloride content in concrete involve meticulous steps to accurately quantify their presence. For sulfate determination, the weight of the ignited crucible containing the sulfate residue is subtracted from the initial weight of the crucible to obtain the weight of the sulfate residue. This weight is then converted to sulfate content using appropriate conversion factors [24].

On the other hand, for chloride analysis, the volume of ammonium thiocyanate solution consumed during titration is recorded. This volume is then multiplied by the normality of the ammonium thiocyanate solution and its equivalent weight to calculate the amount of chloride present in the sample [25].

These calculated values provide crucial insights into the chemical composition of concrete, aiding in assessing its quality, durability, and potential environmental impact. Additionally, these calculations serve as valuable data for engineers and researchers in optimizing concrete mix designs and construction practices to meet specific performance requirements and regulatory standards [26].

Sulfate content:

m_s = Mass of subsample tested
m_1 = Mass of the empty crucible
m_2 = Mass of crucible + precipitate after ignition
$C_{\mathrm{SO}^3} = 34.3\frac{m_2-m_1}{m_s} = \%$ by wt. of concrete.

Chloride content:

$$C_{Cl^-} = \frac{0.3545}{Wt}\left(V\ \mathrm{AgNO_3} - 10\ \mathrm{Ct\ Vt}\right) = \%\ \text{by wt. of concrete.}$$

The results of the sulfate and chloride content analyses offer important insights into the chemical composition of the concrete samples. For sulfate content, calculations based on the mass of the subsample tested, the mass of the empty crucible, and the mass of the crucible plus precipitate after ignition yield percentages by weight of concrete. According to Table 4, the sulfate content of the DRI and L sample is determined to be 0.33%, while for the FTP sample, it is slightly higher at 0.42%. Regarding chloride content, calculations considering the weight tested, volumes of silver nitrate and ammonium thiocyanate solutions used in titration, and respective

Table 4 Sulfate content

Samples.	m_s [g]	m_1[g]	m_2 [g]	m_2 m_1 [g]	C_{SO^3} y t. of concrete
DRI and L	5.00	27.2012	27.2988	0.0476	0.33
FTP	5.002	19.8762	19.9380	0.0618	0.42

Table 5 Chloride content

Samples	W tested [g]	V $AgNO_3$ [mL]	V NH_4SCN (Vt) [mL]	C NH_4SCN(Ct) [mL]	C_{Cl^-}% by wt. of concrete
DRI and L	1.00	25.0	23.4	0.1000	0.57
FTP	1.002	25.0	23.2	0.1000	0.64

concentration factors provide percentages by weight of concrete. As shown in Table 5, the chloride content of the DRI and L sample is measured at 0.57%, whereas in the FTP sample, it is slightly elevated at 0.64%. These results provide valuable data for assessing the quality and durability of the concrete samples, aiding in ensuring compliance with regulatory standards and optimizing construction practices.

Sustainable concrete incorporating sulfur and low chloride and sulfate content represents a significant advancement in the construction industry, offering a range of environmentally friendly alternatives that prioritize both durability and sustainability. In infrastructure projects, such as bridges and highways, where structures are often subjected under harsh environmental conditions and corrosive elements, this innovative concrete formulation stands out for its superior resistance to corrosion [27]. By mitigating the effects of chloride and sulfate exposure, it significantly extends the lifespan of these critical infrastructure elements, reducing the need for frequent repairs and replacements, thereby lowering long-term maintenance costs and minimizing environmental impact.

Furthermore, in the realm of urban construction, where the demand for sustainable building materials is rapidly increasing, this sulfur-containing concrete emerges as a viable solution for eco-conscious high-rise buildings [28]. With its low chemical content, it not only helps minimize environmental degradation during construction but also ensures structural integrity over the building's life cycle. This dual benefit aligns with the growing trend toward green building practices, where developers and architects seek materials that offer both performance and sustainability [29]. In marine environments, such as ports and coastal structures, where the challenges of corrosion and deterioration due to exposure to saltwater are particularly acute, the use of sulfur-containing concrete becomes indispensable [30]. By leveraging sulfur's unique properties, this concrete formulation effectively protects against the damaging effects of chloride and sulfate ions, ensuring the longevity and sustainability of marine infrastructure. This resilience translates to significant cost savings over time, as maintenance and repair efforts are minimized, and the need for premature replacements is reduced.

These diverse applications underscore the versatility and eco-conscious benefits of integrating sustainable concrete into various construction projects. Beyond addressing immediate structural needs, this innovative material contributes to broader sustainability goals by reducing carbon emissions, conserving natural resources, and promoting resilience in the face of changing environmental conditions. As the construction industry continues to prioritize sustainability and resilience, sustainable concrete formulations such as those incorporating sulfur are poised to play an increasingly vital role in shaping the built environment of the future.

4 Conclusion

To conclude, when a reinforced concrete structure experiences degradation from increased loads or changes in function, flexural strengthening is needed in the structure. The paper presents a solution with the use of CFRPs to increase the flexural and shear strength of the structure. A case study of a small office building in Italy showed positive results in structural rehabilitation with the application of CFRPs. Adding one to two carboplate sheets increased the resisting moment by 32–50 kNm at midspan, and 60–80 kNm at supports, ensuring a structurally safe building.

Future work may include further investigation and optimization of CFRP composite strengthening techniques, as well as the study of their performance under different environmental conditions and loading scenarios. Additionally, the durability and long-term behavior of CFRP-strengthened structures could also be studied.

References

1. Wang, X., Han, P., Dong, X., Li, X., Bai, X., He, B., Niu, S., Sun, F.: Influences of high-sulphur fly ash on the properties of lightweight cement-treated materials subjected to sulphate corrosion. Appl. Sci. **10**(15), 5217 (2020)
2. Xiang, D., Weining, L., Xun, L., Qiyi, D., Xiaobing, Y.: The study and applications of modern potline fume treatment plant (FTP). Light Met. **2013**, 775–780 (2016)
3. Zhou, Y., Wang, H., Zhu, R., Wei, G., Dong, K.: Green and efficient recovery of stainless-steel dust by iron-bath direct reduction. Sep. Purif. Technol. **308**, 123005 (2023)
4. Dagher, S., Akhozheya, B., Mohamed, H., El Gamal, M., Slimani, H.: Green concrete from palm trees waste. In: 8th Zero Energy Mass Custom Home (ZEMCH 2021), 26–28 October 2021, Dubai, UAE (2021)
5. Belaïd, F.: How does concrete and cement industry transformation contribute to mitigating climate change challenges? Resour. Conserv. Recycl. Adv. **15**, 200084 (2022)
6. ASTM C 33.: Standard Specification for Concrete Aggregates. Annual Book of ASTM Standards; ASTM International: West Conshohocken, PA, USA (2003)
7. Akhozheya, B., Dagher, S., Alzoubi, Y., Veselinova, I.: Advances in science & engineering technology, carbon fiber reinforced polymer adoption for structural rehabilitation. Meas. Interdiscip. Res. Perspect. **22**, 1–10 (2024)

8. Lazovskaya, I., Jakubowski, S., Hlukhau, D., Lazouski, Y.: Serobeton as a perspective building material. Vestn. Polotskogo Gos. Univ. **8**, 53–57 (2017)
9. Xiang, D., Weining, L., Xun, L., Qiyi, D., Xiaobing, Y.: The study and applications of modern potline fume treatment plant (FTP). Light Met., 775–780 (2013)
10. Direct Reduced Iron.: Determination of Apparent Density and Water Absorption of Hot Briquetted Iron (HBI). BS ISO 15968:2016 (N.d.)
11. Deering, C.E., Madekufamba, M., Lesage, K.L., Marriott, R.A.: The flash point of elemental sulfur: effect of heating rates, hydrogen sulfide, and hydrocarbons. Results Eng. **16**, 100629 (2022)
12. Mohamed, A.M., Gamal, M.M.: A sustainable process for the preparation of sulfur cement for use in public works. Adv. Sustain. Manuf., 127–132 (2011)
13. Yun, J., Jeong, K., Youn, J., Lee, D.: Development of side mold control equipment for producing free-form concrete panels. Buildings. **11**(4), 175 (2021)
14. EN 12390-3:2019.: Testing Hardened Honcrete – Part 3: Compressive Strength of Test Specimens. iTeh Standards: Etobicoke, ON, Canada (2019)
15. Zhang, M., Chen, J., Lv, Y., Wang, D., Ye, J.: Study on the expansion of concrete under attack of sulfate and sulfate–chloride ions. Constr. Build. Mater. **39**, 26–32 (2013)
16. Romadhon, E., Hanif, A.: Analysis of compressive strength of sulfur concrete. J. Mech. Civil Ind. Eng. **3**, 07 (2022)
17. Dhir, R., Paine, K., Ho, N., Poon, C., Etxeberria, M., Brito, J., Tam, V.: Use of recycled and secondary aggregates in concrete: an overview. In: UKIERI Concrete Congress (2011)
18. Holland, N., Nichols, A., Nichols, J.: The Use of Hydrated Lime in Concrete as a Cement Replacement: Effect on Compressive Strength. ASTM Special Technical Publication (2012)
19. Jain, B., Sancheti, G.: Influence of silica fume and iron dust on mechanical properties of concrete. Constr. Build. Mater. **409**, 133910 (2023)
20. Dutta, S.K., Sah, R.: Direct reduced iron: production. In: Encyclopedia of Iron, Steel, and Their Alloys (2016)
21. Ahmad, S., Sajal, W.R., Gulshan, F., Hasan, M., Rhamdhani, M.A.: Thermodynamic analysis of caustic–roasting of electric arc furnace dust. Heliyon. **8**(10), e11031 (2022)
22. Arya, C., Buenfeld, N.R., Newman, J.B.: Assessment of simple methods of determining the free chloride ion content of cement paste. Cem. Concr. Res. **17**(6), 907 (1987)
23. Zuquan, J., Wei, S., Yunsheng, Z., Jinyang, J., Jianzhong, L.: Interaction between sulfate and chloride solution attack of concretes with and without Fly Ash. Cem. Concr. Res. **37**(8), 1223 (2007)
24. Zhong, C., Huang, B.: Deterioration process of cementitious material properties under internal Sulphate attack. Appl. Sci. **13**, 3982 (2023)
25. Sui, S., Wilson, W., Georget, F., Maraghechi, H., Kazemi-Kamyab, H., Sun, W., Scrivener, K.: Quantification methods for chloride binding in Portland cement and limestone systems. Cem. Concr. Res. **125**, 105864 (2019)
26. Jin, Z., Sun, W., Zhang, Y., Jiang, J., Lai, J.: Interaction between sulfate and chloride solution attack of concretes with and without fly ash. Cem. Concr. Res. **37**(8), 1223–1232 (2007)
27. Phuyal, K., Sharma, U., Mahar, J., Mondal, K., Mashal, M.: A sustainable and environmentally friendly concrete for structural applications. Sustain. For. **15**(20) (2023)
28. Jiang, R., Xing, Y., Liu, S., Guo, Y., Guo, B.: Study on the preparation and properties of bridge concrete using low carbon aggregates. Materials. **16**(1) (2022)
29. Xiao, X., Shah, S.P., Xiao, J.: Editorial: low-carbon materials and green construction. Low-Carbon Mater. Green Const. **1**(1) (2023)
30. Santhanam, M., Otieno, M.: Deterioration of concrete in the marine environment. In: Marine Concrete Structures (2016)

Toward Sustainable CO_2 Reduction and Brine Utilization: Investigating Alkaline-Enhanced Solvay Processes

Maisa El Gamal, Ameera F. Mohammad, Aya A. H. Mourad, and Noran Mousa

Abbreviations

Al_2O_3	Aluminum oxide
Ca^{2+}	Calcium
CaO	Calcium oxide
CaC_2	Calcium carbide
CLW	Carbide lime
CO_2	Carbon dioxide
SSR	Stainless steel jacketed column reactor
K^+	Potassium
KOH	Potassium hydroxide
Mg^{2+}	Magnesium
Na^+	Sodium
NaCl	Sodium chloride
$NaHCO_3$	Sodium bicarbonate
NH_3	Ammonia
NH_4OH	Ammonium hydroxide
NH_4Cl	Ammonium chloride

M. El Gamal (✉)
College of Natural and Health Sciences, Zayed University, Abu Dhabi, UAE
e-mail: maisa.elgamal@zu.ac.ae

A. F. Mohammad
Engineering Requirements Unit, UAE University, Al Ain, UAE

A. A. H. Mourad
Abu Dhabi Polytechnic, Institute of Applied Technology, Abu Dhabi, UAE

N. Mousa
Department of Chemical and Petroleum Engineering, UAE University, Al Ain, UAE

L. Chen (ed.), *Trends in Clean Energy Research*, Green Energy and Technology,
https://doi.org/10.1007/978-3-031-67987-2_25

NH_4HCO_3	Ammonium bicarbonate
UAE	United Arab Emirates

1 Introduction

The escalating threat of global warming poses a significant challenge to modern industrial societies, primarily driven by the emission of greenhouse gases such as carbon dioxide, nitrous oxide, and methane into the atmosphere. Carbon dioxide, with its predominant presence accounting for approximately 60% of greenhouse gas emissions [1], stands as the chief contributor to the warming effect, directly impacting the Earth's surface temperature. International agreements aim to mitigate global warming and climate change by halving CO_2 emissions by 2050, limiting the rise in Earth's surface temperature to 2 °C or less. To achieve this target, CO_2 emissions must be reduced by at least 25% by 2030 [2]. CO_2 capture from fossil-fueled power plants presents a potential solution to control greenhouse gas emissions, as these plants account for about 40% of total CO_2 emissions [3]. However, existing CO_2 capture methods, including pre-combustion, post-combustion, oxyfuel, and chemical looping combustion, face limitations hindering large-scale industrial implementation [4]. Notably, post-combustion CO_2 capture, widely adopted in new fossil fuel-burning power plants, encompasses physical, chemical, and biological methods, each with distinctive approaches [5].

On the other hand, potable water scarcity is a pressing global issue, exacerbated by population growth surpassing conventional water resources. With over 1 billion lacking access to clean drinking water and approximately 2.3 billion facing water shortages, the situation is dire [6]. Projections indicate impending severe water scarcity in the European Union and northern Mediterranean countries by 2030 [7]. The UAE exemplifies this challenge, grappling with water shortages due to population growth and expanding industrial and agricultural activities [8]. Desalination has emerged as a vital solution, with thermal and membrane methods evolving over decades [9]. The Middle East leads in seawater desalination, boasting around 50% of global production capacity despite having just 2.9% of the world's population [10]. Key desalination technologies include reverse osmosis and distillation methods such as multistage flash and vapor compression distillation. The reject brine, subjected to analysis in this study, is a high saline water by-product resulting from these processes. Typically, reject brine is disposed of due to its elevated salinity levels, posing challenges for safe discharge or reuse. Efficient management strategies are imperative to address the environmental impacts associated with its disposal.

There is an urgent imperative to devise a novel approach for managing desalination reject brine, given its underutilization and overlooked environmental ramifications. Utilizing carbon dioxide for chemical reactions with rejected brine presents a promising, cost-effective, and eco-friendly solution [11]. Employing the Solvay process, rejected brine can be converted into sodium bicarbonate, a valuable solid product, while the treated water finds applications in irrigation and industry.

Additionally, this method facilitates CO_2 capture from industrial exhaust or flue gases. Studies indicate that optimizing CO_2 reactions with ammoniated brine at 20 °C yields favorable conversions using various CO_2 forms [11].

The Solvay process, named after Ernest Solvay, originated in 1881 for sodium carbonate production. It involves reacting concentrated brine with ammonia (NH_3) and carbon dioxide to produce soluble NH_4HCO_3, which then reacts with NaCl to yield soluble NH_4Cl and $NaHCO_3$ precipitate [12]. The NH_4Cl can further react with calcium hydroxide to regenerate ammonia. Although ammonia does not partake in the overall reaction, it influences intermediate reactions and aids in precipitating sodium bicarbonate. Maximizing sodium bicarbonate conversion requires minimizing its solubility, highlighting the importance of optimizing relevant factors [12].

In this study, various alkaline materials, alongside ammonia, will be explored to assess their influence on brine desalination by precipitating carbonated and bicarbonated products via traditional and modified Solvay processes. The reduction in salinity resulting from this process will be analyzed, along with its impact on the efficiency of CO_2 capture from simulated effluent gases.

2 Methodology and Experimental Apparatus

The experimental setup revolves around a specially designed SSR with dimensions of 78 mm internal diameter and 700 mm overall height. Operating in semi-batch mode, the reactor exposes the liquid to a continuous flow of a carbon dioxide-air mixture, with temperature control facilitated by water bath circulation through the jacket. Gas input at the reactor's bottom is regulated by a gas flow controller, while liquid input at the top is managed by a piston pump. The reactor features a sampling port for liquid and bottom discharge capability. Gas effluent exits the top, passing through a moisture trap before analysis via a CO_2 gas analyzer. Figure 1 depicts a schematic diagram of the SSR system. Reject brine samples, with salinity levels ranging between 65,000 and 75,000 ppm, were sourced from a local desalination plant utilizing the RO desalination process. The brine's characteristics are detailed in Table 1. NH_4OH (25 wt. % NH_3), CaO, KOH, and Al_2O_3 utilized in the study were acquired from Scientific Progress Medical and Scientific Equipment Company in the UAE. CaO and Al_2O_3 boast a purity level of 99.9%, while KOH exhibits a purity range of 85–100%. Carbide lime samples were collected from Arabian Industrial Gases Co., Ras Al-Khaimah, UAE. A gas mixture comprising 10% CO_2 and 90% air was sourced from Abu Dhabi Oxygen Company, UAE.

The SSR system was initially charged with the prepared solution and subjected to a continuous gas flow containing 10% volume of CO_2. To quantify the amount of CO_2 captured, the gas exiting the reactor was directed through a CO_2 gas analyzer. The internal pressure within the reactor was carefully managed using a back pressure regulator valve installed at the outlet to maintain optimal operating conditions. Each experimental run was conducted until reaching the designated

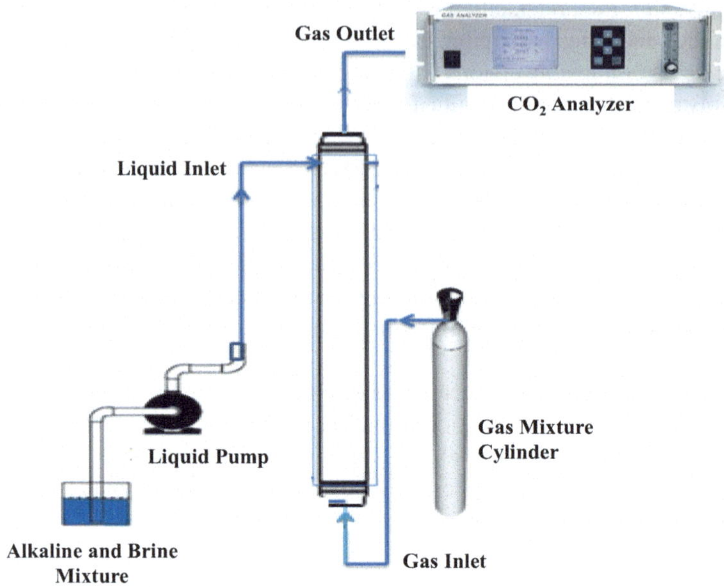

Fig. 1 A schematic diagram of the stainless steel column reactor (SSR)

Table 1 Characteristics of the reject brine

Parameter	Value
pH	9.16
TDS	73.8 g/l
Salinity	71,700 ppm
COD	1560 ppm
Na^+	23,712 ppm
Mg^{2+}	2794 ppm
K^+	762 ppm
Ca^{2+}	1375 ppm

saturation point, defined as the stage where the CO_2 concentration in the outlet gas stream equaled its inlet value of 10% volume. Throughout the experiment, liquid samples were extracted at 15-min intervals, and the pH value of each sample was meticulously recorded to monitor changes in solution chemistry over time. Upon completion of each experiment, the contents of the reactor were systematically drained, and the treated brine liquid was separated from any precipitated solids using vacuum filtration facilitated by a Buchner funnel setup. The ion concentrations of the treated brine were subsequently analyzed using inductively coupled plasma spectrometry. Key performance indicators, including the percentage reduction of sodium concentration, moles of CO_2 reacted, and CO_2 uptake, were calculated using specific equations. The moles of CO_2 reacted were computed using Eq. (1). Additionally, the CO_2 uptake was calculated using Eq. (2):

$$\text{Moles of } CO_2 \text{ reacted} = \frac{\int_0^t \text{volume of } CO_2\text{captured } (L/\min)\,.dt}{\text{Molar volume of } CO_2 \ (L/mol)} \tag{1}$$

$$CO_2 \text{ uptake} = \frac{\text{Mass of } CO_2 \text{ reacted}}{\text{Mass of KOH used}} \tag{2}$$

3 Results and Discussions

The traditional Solvay process was implemented for CO_2 capture and brine desalination, the method involving the reaction of carbon dioxide with ammoniated brine to yield $NaHCO_3$ and soluble NH_4Cl, thereby concurrently reducing sodium concentration in reject brine and mitigating carbon dioxide emissions. Through response surface methodology, optimal conditions for maximizing CO_2 capture efficiency and ions removal were determined, as a process conducted at a temperature of approximately 20 °C, a gas flow rate of 1 L/min, and a molar ratio of 3.3 NH_3:1NaCl (158.4 g NH_4OH). Under these optimized conditions, a CO_2 capture uptake value of 0.292 g CO_2/g NH_4OH accompanied by a maximum sodium removal rate of 33% was achieved. Figure 2 illustrates the CO_2 capture efficiency percentage for the modified Solvay process based on NH_4OH over 300 min. These findings underscore the technical viability of employing the Solvay approach for CO_2 capture via reactions with desalination reject brine [13].

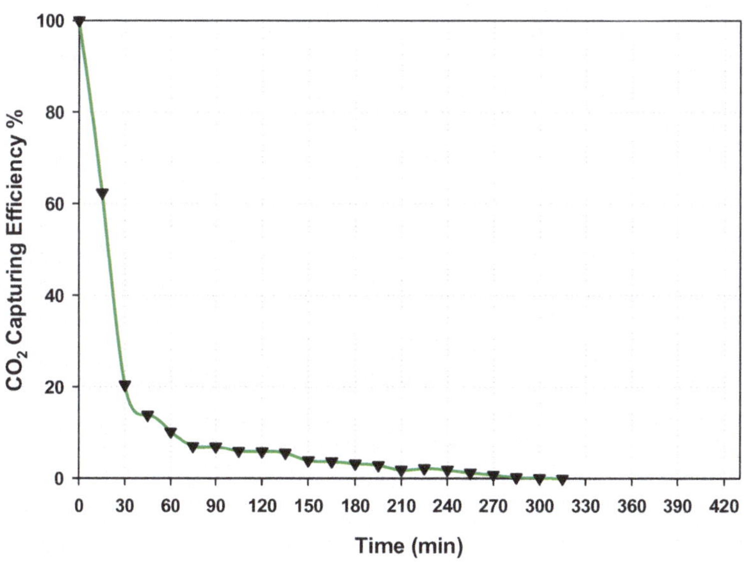

Fig. 2 CO_2 capture efficiency at 158.4 g NH4OH/l (excess amount of NH4OH) versus reaction time with 10% of CO_2, a gas flow rate of 1 L/min, and a temperature of 20 °C

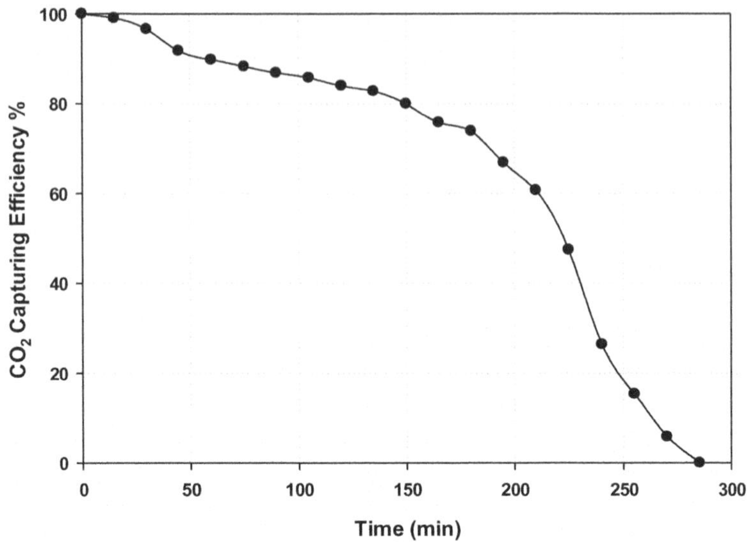

Fig. 3 CO_2 capture efficiency at 110 g/l of KOH versus reaction time with 10% of CO_2, a gas flow rate of 848.5 mL/min, and a temperature of 10 °C

In another investigation, the effectiveness of a modified Solvay process was explored under conditions where ammonia was absent. This involved simultaneous treatments of CO_2 and brine in the presence of KOH (110 g/l brine) and Al_2O_3 (21.4 g/l brine). The effect of these two alkaline materials on key parameters such as pH levels, CO_2 capture efficiency, ion reduction, and the nature of generated solids was evaluated. The KOH/brine mixture achieved a CO_2 uptake of 0.58 g CO_2/g KOH under optimum conditions of 10 °C and gas flow rate of 848.5 mL/min, and the data are demonstrated in Fig. 3. Furthermore, the percentage reduction of ions for KOH was found to reach 44.1% sodium reduction. Conversely, Al_2O_3 demonstrated the poorest performance in CO_2 capture with an uptake value of 0.013 g CO_2/g Al_2O_3 and ion removal of 2.03% sodium reduction under 20 °C and 1 L/min gas flow rate. Figure 4 shows the CO_2 capturing efficiency percentage for the modified Solvay process based on Al_2O_3 material. The results confirm the utilization of KOH as a promising alkaline agent for the combined process. Notably, the utilization of KOH resulted in the production of potassium chloride crystals, a valuable by-product that can be easily separated. While Al_2O_3 displayed minimal reactivity, it demonstrated efficacy in recovering magnesium ions from the reject brine solution with 24.0% magnesium recovery, suggesting potential application as a coagulant for this purpose [14].

In different levels of Solvay process modification and eliminating the use of ammonia, a new process involves the direct reaction of CO_2 with saline water in the presence of CaO (20 g CaO/l brine). Detailed investigations were conducted to analyze the impact of various operating parameters, including reaction temperature, water pH, and stoichiometry, on both CO_2 capture efficiency and sodium

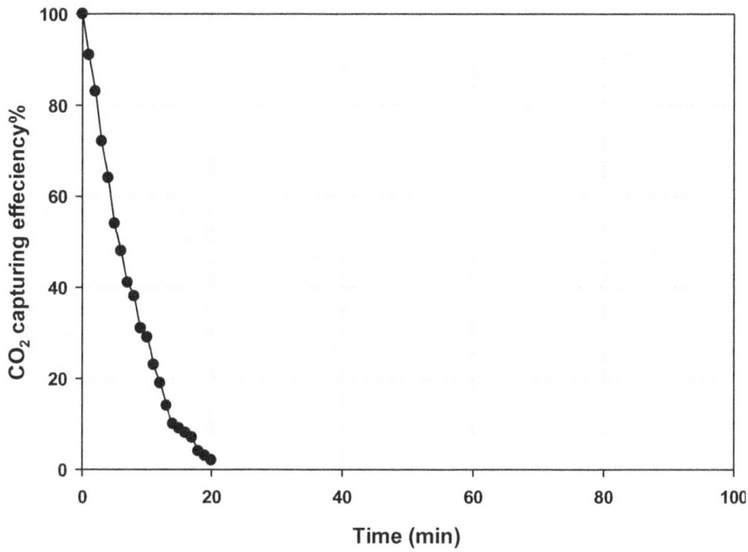

Fig. 4 CO_2 capture efficiency at 21.4 g/l of Al_2O_3 (stoichiometry amount) versus reaction time with 10% of CO_2, a gas flow rate of 1 L/min, and a temperature of 20 °C

removal efficiency. Through response surface methodology, optimal conditions for maximizing CO_2 capture efficiency and sodium removal were determined, revealing that a temperature of 20 °C and a pH exceeding 10 were most conducive for the modified process. Under these optimized conditions, the modified process using CaO surpassed the traditional Solvay process with a capture uptake of 0.921 g CO_2/g CaO and sodium removal rates of 35%. Figure 5 demonstrates the CO_2 capture efficiency percentage for the modified Solvay process based on CaO. Notably, the pH emerged as a critical determinant of process efficacy, with higher pH levels correlating with enhanced performance in both CO_2 capture and sodium removal. Overall, experimental findings underscored the superiority of the modified Solvay process utilizing CaO as the alkaline material in terms of CO_2 capture efficiency, sodium removal, and energy consumption [15].

An innovative method for CO_2 capture and brine desalination diverges from the conventional Solvay process by eliminating the need for ammonia. Instead, this approach involves the direct reaction of CO_2 with reject brine, in the presence of CLW (100 g CLW/l brine), also known as lime sludge. CLW is a by-product generated during the synthesis of acetylene through the hydrolysis reaction of CaC_2. Extensive investigations were carried out to assess the influence of various operational parameters, including reaction temperature, water pH, and stoichiometry, on both CO_2 capture efficiency and sodium removal efficiency in both the traditional and modified Solvay processes. Utilizing response surface methodology, optimal conditions were identified to maximize CO_2 capture efficiency and sodium removal. These investigations revealed that a temperature of 20 °C and a pH exceeding 10 were most favorable for the new process. Under these optimized conditions, the CO_2

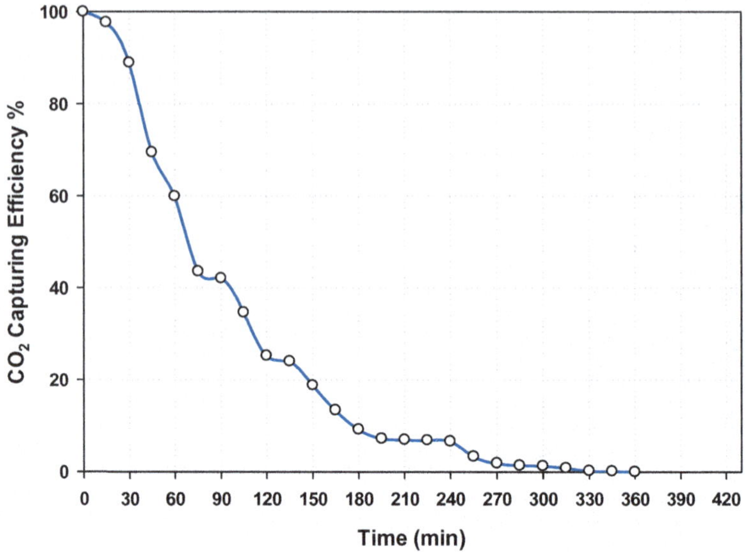

Fig. 5 CO_2 capture efficiency at 20.0 g/l (excess amount of CaO) versus reaction time with 10% of CO_2, a gas flow rate of 1 L/min, and a temperature of 20 °C

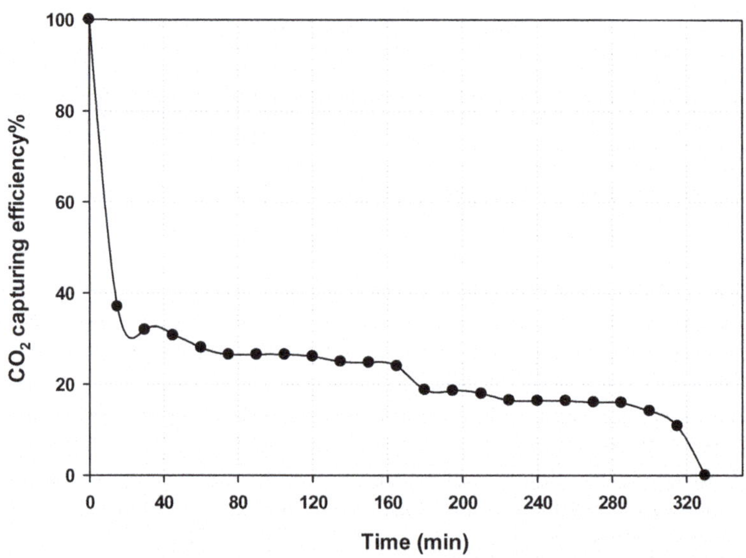

Fig. 6 CO_2 capture efficiency at 100 g carbide lime/l (excess amount of carbide lime) versus reaction time with 10% of CO_2, a gas flow rate of 1 L/min, and a temperature of 20 °C

uptake value reached 0.1779 g CO_2/g CLW, with sodium removal rates achieving 47.1% [16]. Figure 6 shows the CO_2 capturing percentage for the modified Solvay process based on carbide lime waste.

Table 2 CO_2 uptake values per unit mass of alkaline material used in various Solvay processes

Type of alkaline material used	CO_2 uptake value/mass of alkaline material used
KOH	0.58 g CO_2/g KOH (110 g KOH/l brine)
CaO	0.921 g CO_2/g CaO (20 g CaO/l brine)
NH_4OH	0.292 g CO_2/gNH_4OH (158.4 g NH_4OH/l brine)
CLW	0.1779 g CO_2/g CLW (100 g CLW/l brine)
Al_2O_3	0.013 g CO_2/g Al_2O_3 (20 g Al_2O_3/l brine)

Table 2 provides a comprehensive overview of the performance of the aforementioned processes in capturing CO_2, quantified as the uptake value per unit mass of the alkaline material utilized. It becomes evident that the modified Solvay processes utilizing KOH and CaO exhibit significantly higher CO_2 uptake values compared to the traditional Solvay process employing NH_4OH. This enhancement in CO_2 capture efficiency underscores the efficacy of the modified methodologies in achieving greater CO_2 sequestration per unit mass of alkaline material utilized.

4 Conclusions

The traditional Solvay process, renowned for CO_2 capture and brine desalination, underwent scrutiny in a semi-batch reactor. Utilizing response surface methodology, conditions were optimized, resulting in a CO_2 uptake of 0.292 g CO_2/g NH_4OH over 300 min, with a maximum sodium removal rate of 33%. A modified Solvay process, devoid of ammonia, employed KOH and Al_2O_3. Optimization yielded significant CO_2 uptake (0.58 g CO_2/g KOH) and sodium reduction (44.1%). Moreover, a novel approach involving the direct reaction of CO_2 with saline water alongside CaO surpassed conventional methods, achieving superior CO_2 capture efficiency (0.921 g CO_2/g CaO) and sodium removal rates (35%). Further innovation entailed utilizing carbide lime wastes (CLW), resulting in a CO_2 uptake value of 0.1779 g CO_2/g CLW and a sodium removal rate of 47.1% under favorable conditions of 20 °C and a pH exceeding 10.

References

1. Hashimoto, K., et al.: Global CO_2 recycling – novel materials and prospect for prevention of global warming and abundant energy supply. Mater. Sci. Eng. A. **267**(2), 200–206 (1999)
2. den Elzen, M., Meinshausen, M., van Vuuren, D.: Multi-gas emission envelopes to meet greenhouse gas concentration targets: costs versus certainty of limiting temperature increase. Glob. Environ. Change. **17**(2), 260–280 (2007)
3. Holtz-Eakin, D., Selden, T.M.: Stoking the fires? CO_2 emissions and economic growth. J. Public Econ. **57**(1), 85–101 (1995)

4. Kunze, C., Spliethoff, H.: Assessment of oxy-fuel, pre-and post-combustion-based carbon capture for future IGCC plants. Appl. Energy. **94**, 109–116 (2012)
5. Bryngelsson, M., Westermark, M.: CO_2 capture pilot test at a pressurized coal fired CHP plant. Energy Procedia. **1**(1), 1403–1410 (2009)
6. Morillo, J., Usero, J., Rosado, D., El Bakouri, H., Riaza, A., Bernaola, F.-J.: Comparative study of brine management technologies for desalination plants. Desalination. **336**, 32–49 (2014)
7. Le Dirach, J., Nisan, S., Poletiko, C.: Extraction of strategic materials from the concentrated brine rejected by integrated nuclear desalination systems. Desalination. **182**(1–3), 449–460 (2005)
8. Mohamed, A., Maraqa, M., Al Handhaly, J.: Impact of land disposal of reject brine from desalination plants on soil and groundwater. Desalination. **182**(1–3), 411–433 (2005)
9. Greenlee, L.F., Lawler, D.F., Freeman, B.D., Marrot, B., Moulin, P.: Reverse osmosis desalination: water sources, technology, and today's challenges. Water Res. **43**(9), 2317–2348 (2009)
10. Sauvet-Goichon, B.: Ashkelon desalination plant – a successful challenge. Desalination. **203**(1–3), 75–81 (2007)
11. El-Naas, M.H., Al-Marzouqi, A.H., Chaalal, O.: A combined approach for the management of desalination reject brine and capture of CO_2. Desalination. **251**(1–3), 70–74 (2010)
12. El-Naas, M.H.: Reject brine management. In: Desalination, Trends and Technologies, pp. 237–252 (2011)
13. Mohammad, A.F., El-Naas, M.H., Suleiman, M.I., Al Musharfy, M.: Optimization of a solvay-based approach for CO_2 capture. Int. J. Chem. Eng. Appl. **7**(4), 230 (2016)
14. Mourad, A.A.H., Mohammad, A.F., Altarawneh, M., Al-Marzouqi, A.H., El-Naas, M.H., Al-Marzouqi, M.H.: Effects of potassium hydroxide and aluminum oxide on the performance of a modified solvay process for CO_2 capture: A comparative study. Int. J. Energy Res. **45**(9), 13952–13964 (2021)
15. El-Naas, M.H., Mohammad, A.F., Suleiman, M.I., Al Musharfy, M., Al-Marzouqi, A.H.J.D.: A new process for the capture of CO_2 and reduction of water salinity. Desalination. **411**, 69–75 (2017)
16. El Gamal, M., Mohammad, A., Hameedi, S., AlZawahreh, H.J.E.R.: Carbon dioxide uptakes by acetylene by-products through gas–solid and gas–solid–liquid reactions. Energy Rep. **7**, 193–198 (2021)

Correction to: State of Decision-Making in the Baltic States: Nuclear Energy Past and for the Future

Guna Jakobsone-Snepste, Guillaume Deffuant, Ieva Vitolina, Janis Kapenieks Sr., and Valdis Vitolins

Correction to: Chapter 14 in: L. Chen (ed.), *Trends in Clean Energy Research*, Green Energy and Technology, https://doi.org/10.1007/978-3-031-67987-2_14

The chapter "State of Decision-Making in the Baltic States: Nuclear Energy Past and for the Future", written by Guna Jakobsone-Snepste, Guillaume Deffuant, Ieva Vitolina, Janis Kapenieks Sr., and Valdis Vitolins, was originally published without open access. Following the authors' decision to opt for open access, the copyright of the chapter changed on 17th March 2025 to © The Author(s) 2025 and the chapter is now distributed under the terms of the Creative Commons Attribution 4.0 International License (https://creativecommons.org/licenses/by/4.0/), which permits use, sharing, adaptation, distribution and reproduction in any medium or format, as long as you give appropriate credit to the original author(s) and the source, provide a link to the Creative Commons licence and indicate if changes were made.

The images or other third party material in this chapter are included in the chapter's Creative Commons licence, unless indicated otherwise in a credit line to the material. If material is not included in the chapter's Creative Commons licence and your intended use is not permitted by statutory regulation or exceeds the permitted use, you will need to obtain permission directly from the copyright holder.

In addition, the author names of this chapter were incorrectly cited. These should have been cited as Guna Jakobsone-Snepste, Guillaume Deffuant, Ieva Vitolina, Janis Kapenieks Sr., and Valdis Vitolins.

The updated version of this chapter can be found at
https://doi.org/10.1007/978-3-031-67987-2_14

L. Chen (ed.), *Trends in Clean Energy Research*, Green Energy and Technology,
https://doi.org/10.1007/978-3-031-67987-2_26

Index

L. Chen (ed.), *Trends in Clean Energy Research*, Green Energy and Technology,
https://doi.org/10.1007/978-3-031-67987-2

The manufacturer's authorised representative in the EU is Springer Nature Customer Service Centre GmbH, Europaplatz 3, 69115 Heidelberg, Germany. If you have any concerns regarding our products, please contact ProductSafety@springernature.com

Printed and bound by CPI Group (UK) Ltd, Croydon, CR0 4YY

01/10/2026

02225906-0004